Mining: Processes, Tools and Techniques

Mining: Processes, Tools and Techniques

Edited by
Diana Bates

Larsen & Keller
www.larsen-keller.com

Mining: Processes, Tools and Techniques
Edited by Diana Bates
ISBN: 978-1-63549-186-9 (Hardback)

© 2017 Larsen & Keller

☰ Larsen & Keller

Published by Larsen and Keller Education,
5 Penn Plaza,
19th Floor,
New York, NY 10001, USA

Cataloging-in-Publication Data

Mining : processes, tools and techniques / edited by Diana Bates.
     p. cm.
Includes bibliographical references and index.
ISBN 978-1-63549-186-9
1. Mines and mineral resources. 2. Mining engineering.
3. Mineral industries. I. Bates, Diana.
TN153 .M56 2017
622--dc23

The publisher's policy is to use permanent paper from mills that operate a sustainable forestry policy. Furthermore, the publisher ensures that the text paper and cover boards used have met acceptable environmental accreditation standards.

Printed and bound in the United States of America.

For more information regarding Larsen and Keller Education and its products, please visit the publisher's website www.larsen-keller.com

Table of Contents

Preface

Mining refers to the practice of extracting minerals that are present inside and on earth. Mining results in the extraction of ores like limestones, metals, clay, gemstone, gravel, coal, etc. It is also used to extract vital non-renewable resources like petroleum, water and natural gas. It is an important activity which results in economic wealth for many countries and also provides us with crucial elements. This book is compiled in such a manner, that it will provide in-depth knowledge about the theory and practice of mining. Some of the diverse topics covered in it address the varied branches that fall under this category. Most of the topics introduced in this book covers new techniques and applications of mining. The book will serve as a reference to a broad spectrum of readers.

To facilitate a deeper understanding of the contents of this book a short introduction of every chapter is written below:

Chapter 1- The extraction of valuable minerals or other geological materials from the earth is called mining. Mining has been practiced for decades and is one of the main sources of obtaining minerals. The chapter on mining offers an insightful focus, keeping in mind the complex subject matter.

Chapter 2- Mining is practiced in many countries and can be grouped into major categories such as coal mining, gold mining, metal ore mining and placer mining. Mining is best understood in confluence with the major topics listed in the following chapter. The topics discussed in the chapter are of great importance to broaden the existing knowledge on mining.

Chapter 3- The methods and processes involved in any field of study is an important component of the study. This chapter involves the reader with a better understanding on automated mining, asteroid mining, deep-sea mining, Uranium mining and bio mining. The chapter serves as a source to understand the major categories related to mining.

Chapter 4- Surface mining is a part of mining. In contrast to underground mining, surface mining is the removal of soil and rock overlying the mineral deposit. The techniques explained in this chapter are open pit mining, mountaintop removal mining, sand mining and dredging.

Chapter 5- Underground mining can best be understood in regard to drift mining, stoping, shaft mining and drilling and blasting and other techniques. Stoping is the extraction of the desired ore or any other mineral from an underground mine, while shaft mining is excavating a tunnel from the top down where there is initially no access to the bottom. This chapter develops a profound understanding in the reader, related to underground mining.

Chapter 6- The major components discussed in this chapter are coal, oil shale and clay. This chapter helps the reader to develop a comprehensive understanding on the minerals extracted through mining. The aspects elucidated are of vital importance and provide a better understanding on mining.

Chapter 7- Mining has adverse effects on the environment, and protective measures need to be taken in regard to this. The environmental effect of mining includes erosion, loss of biodiversity and contamination of ground water and surface water. The topics discussed in the chapter are of great importance to broaden the existing knowledge on the impacts of mining on the environment.

Finally, I would like to thank the entire team involved since the inception of this book for their valuable time and contribution. This book would not have been possible without their efforts. I would also like to thank my friends and family for their constant support.

Editor

Introduction to Mining

The extraction of valuable minerals or other geological materials from the earth is called mining. Mining has been practiced for decades and is one of the main sources of obtaining minerals. The chapter on mining offers an insightful focus, keeping in mind the complex subject matter.

Mining is the extraction of valuable minerals or other geological materials from the earth from an orebody, lode, vein, seam, reef or placer deposits which forms the mineralized package of economic interest to the miner.

Surface coal mining

Ores recovered by mining include metals, coal, oil shale, gemstones, limestone, dimension stone, rock salt, potash, gravel, and clay. Mining is required to obtain any material that cannot be grown through agricultural processes, or created artificially in a laboratory or factory. Mining in a wider sense includes extraction of any non-renewable resource such as petroleum, natural gas, or even water.

Mining of stones and metal has been a human activity since pre-historic times. Modern mining processes involve prospecting for ore bodies, analysis of the profit potential of a proposed mine, extraction of the desired materials, and final reclamation of the land after the mine is closed.

Mining operations usually create a negative environmental impact, both during the mining activity and after the mine has closed. Hence, most of the world's nations have passed regulations to decrease the impact. Worker safety has long been a concern as well, and modern practices have significantly improved safety in mines.

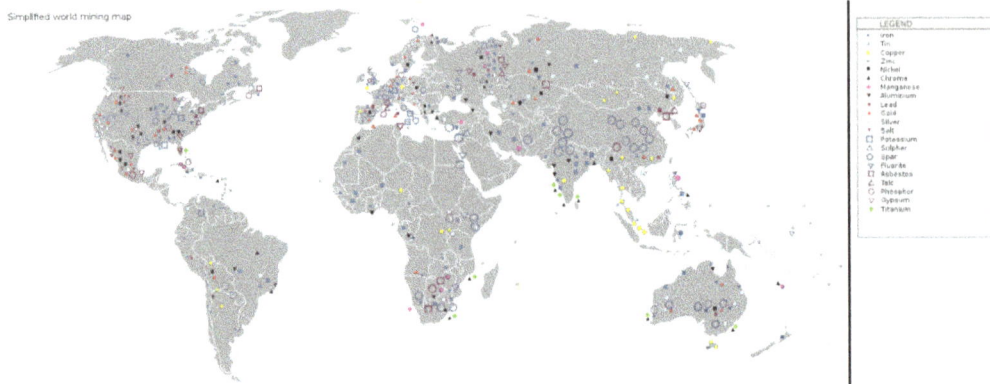

Simplified world active active mining map

Levels of metals recycling are generally low. Unless future end-of-life recycling rates are stepped up, some rare metals may become unavailable for use in a variety of consumer products. Due to the low recycling rates, some landfills now contain higher concentrations of metal than mines themselves.

Sulfur miner with 90 kg of sulfur carried from the floor of the Ijen Volcano (2015)

History

Prehistoric Mining

Since the beginning of civilization, people have used stone, ceramics and, later, metals found close to the Earth's surface. These were used to make early tools and weapons; for example, high quality flint found in northern France and southern England was used to create flint tools. Flint mines have been found in chalk areas where seams of the stone were followed underground by shafts and galleries. The mines at Grimes Graves are especially famous, and like most other flint mines, are Neolithic in origin (ca 4000 BC-ca 3000 BC). Other hard rocks mined or collected for axes included the greenstone of the Langdale axe industry based in the English Lake District.

The oldest known mine on archaeological record is the "Lion Cave" in Swaziland, which radiocarbon dating shows to be about 43,000 years old. At this site Paleolithic humans mined hematite to make the red pigment ochre. Mines of a similar age in Hungary are believed to be sites where Neanderthals may have mined flint for weapons and tools.

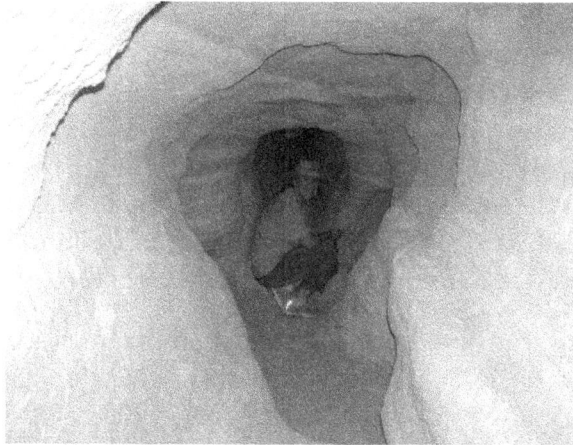
Chalcolithic copper mine in Timna Valley, Negev Desert

Ancient Egypt

Ancient Egyptians mined malachite at Maadi. At first, Egyptians used the bright green malachite stones for ornamentations and pottery. Later, between 2613 and 2494 BC, large building projects required expeditions abroad to the area of Wadi Maghareh in order to secure minerals and other resources not available in Egypt itself. Quarries for turquoise and copper were also found at Wadi Hamamat, Tura, Aswan and various other Nubian sites on the Sinai Peninsula and at Timna.

Mining in Egypt occurred in the earliest dynasties. The gold mines of Nubia were among the largest and most extensive of any in Ancient Egypt. These mines are described by the Greek author Diodorus Siculus, who mentions fire-setting as one method used to break down the hard rock holding the gold. One of the complexes is shown in one of the earliest known maps. The miners crushed the ore and ground it to a fine powder before washing the powder for the gold dust.

Ancient Greek and Roman Mining

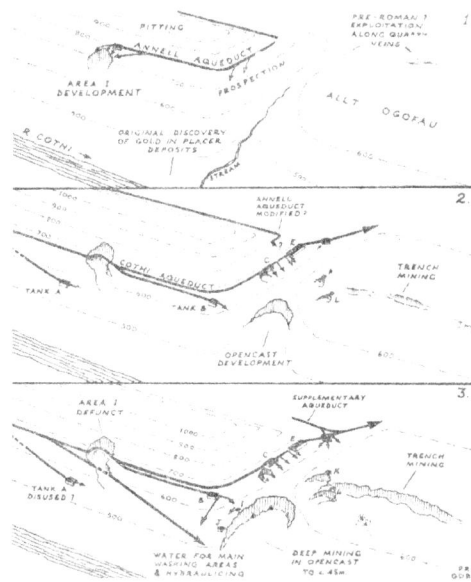
Ancient Roman development of the Dolaucothi Gold Mines, Wales

Mining in Europe has a very long history. Examples include the silver mines of Laurium, which helped support the Greek city state of Athens. Despite the mine having over 20,000 slaves working in them, the technology was essentially identical to their Bronze Age predecessors. Other mines, such as on the island of Thassos, had marble quarried by the Parians after having arrived in the 7th Century BC. The marble was shipped away and was later found by archaeologists to have been used in buildings including the tomb of Amphipolis. Philip II of Macedon, the father of Alexander the Great, captured the gold mines of Mount Pangeo in 357 BC to fund his military campaigns. He also captured gold mines in Thrace for minting coinage, eventually producing 26 tons per year.

However, it was the Romans who developed large scale mining methods, especially the use of large volumes of water brought to the minehead by numerous aqueducts. The water was used for a variety of purposes, including removing overburden and rock debris, called hydraulic mining, as well as washing comminuted, or crushed, ores and driving simple machinery.

The Romans used hydraulic mining methods on a large scale to prospect for the veins of ore, especially a now obsolete form of mining known as hushing. This method involved building numerous aqueducts to supply water to the minehead where it was stored in large reservoirs and tanks. When a full tank was opened, the flood of water sluiced away the overburden to expose the bedrock underneath and any gold veins. The rock was then worked upon by fire-setting to heat the rock, which would be quenched with a stream of water. The resulting thermal shock cracked the rock, enabling it to be removed, aided by further streams of water from the overhead tanks. The Roman miners used similar methods to work cassiterite deposits in Cornwall and lead ore in the Pennines.

The methods had been developed by the Romans in Spain in 25 AD to exploit large alluvial gold deposits, the largest site being at Las Medulas, where seven long aqueducts were built to tap local rivers and to sluice the deposits. Spain was one of the most important mining regions, but all regions of the Roman Empire were exploited. In Great Britain the natives had mined minerals for millennia, but after the Roman conquest, the scale of the operations increased dramatically, as the Romans needed Britannia's resources, especially gold, silver, tin, and lead.

Roman techniques were not limited to surface mining. They followed the ore veins underground once opencast mining was no longer feasible. At Dolaucothi they stoped out the veins, and drove adits through barren rock to drain the stopes. The same adits were also used to ventilate the workings, especially important when fire-setting was used. At other parts of the site, they penetrated the water table and dewatered the mines using several kinds of machines, especially reverse overshot water-wheels. These were used extensively in the copper mines at Rio Tinto in Spain, where one sequence comprised 16 such wheels arranged in pairs, and lifting water about 80 feet (24 m). They were worked as treadmills with miners standing on the top slats. Many examples of such devices have been found in old Roman mines and some examples are now preserved in the British Museum and the National Museum of Wales.

Medieval Europe

Mining as an industry underwent dramatic changes in medieval Europe. The mining industry in the early Middle Ages was mainly focused on the extraction of copper and iron. Other precious metals were also used, mainly for gilding or coinage. Initially, many metals were obtained through open-pit mining, and ore was primarily extracted from shallow depths, rather than through deep mine shafts. Around the 14th century, the growing use of weapons, armour, stirrups, and horse-shoes greatly increased the demand for iron. Medieval knights, for example, were often laden with up to 100 pounds of plate or chain link armour in addition to swords, lances and other weapons. The overwhelming dependency on iron for military purposes spurred iron production and extraction processes.

Agricola, author of *De Re Metallica*

The silver crisis of 1465 occurred when all mines had reached depths at which the shafts could no longer be pumped dry with the available technology. Although an increased use of bank notes, credit and copper coins during this period did decrease the value of, and dependence on, precious metals, gold and silver still remained vital to the story of medieval mining.

Gallery, 12th to 13th century, Germany

Due to differences in the social structure of society, the increasing extraction of mineral deposits spread from central Europe to England in the mid-sixteenth century. On the continent, all mineral deposits belonged to the crown, and this regalian right was stoutly maintained; but in England, it was pared down to gold and silver (of which there were virtually no deposits) by a judicial decision of 1568 and a law of 1688. England had iron, zinc, copper, lead, and tin ores. Landlords who owned the base metals and coal under their estates were now rendered with a strong inducement to extract these metals or to lease the deposits and collect royalties from mine operators. English,

German, and Dutch capital combined to finance extraction and refining. Hundreds of German technicians and skilled workers were brought over; in 1642 a colony of 4,000 foreigners was mining and smelting copper at Keswick in the northwestern mountains.

Use of water power in the form of water mills was extensive. The water mills were employed in crushing ore, raising ore from shafts, and ventilating galleries by powering giant bellows. Black powder was first used in mining in Selmecbánya, Kingdom of Hungary (now Banská Štiavnica, Slovakia) in 1627. Black powder allowed blasting of rock and earth to loosen and reveal ore veins. Blasting was much faster than fire-setting and allowed the mining of previously impenetrable metals and ores. In 1762, the world's first mining academy was established in the same town.

The widespread adoption of agricultural innovations such as the iron plowshare, as well as the growing use of metal as a building material, was also a driving force in the tremendous growth of the iron industry during this period. Inventions like the arrastra were often used by the Spanish to pulverize ore after being mined. This device was powered by animals and used the same principles used for grain threshing.

Much of the knowledge of medieval mining techniques comes from books such as Biringuccio's *De la pirotechnia* and probably most importantly from Georg Agricola's *De re metallica* (1556). These books detail many different mining methods used in German and Saxon mines. One of the prime issues confronting medieval miners (and one which Agricola explains in detail) was the removal of water from mining shafts. As miners dug deeper to access new veins, flooding became a very real obstacle. The mining industry became dramatically more efficient and prosperous with the invention of mechanical and animal driven pumps.

Classical Philippine Civilization

The image of a Maharlika class of the Philippine Society , depicted in Boxer Codex that the Gold used as a form of Jewelry (ca.1400).

Mining in the Philippines began around 1000 BC. The early Filipinos worked various mines of gold, silver, copper and iron. Jewels, gold ingots, chains, calombigas and earrings were handed down from antiquity and inherited from their ancestors. Gold dagger handles, gold dishes, tooth plating, and huge gold ornamets were also used. In Laszlo Legeza's "Tantric elements in pre-Hispanic Philippines Gold Art", he mentioned that gold jewelry of Philippine origin was found in Ancient Egypt. According to Antonio Pigafetta, the people of Mindoro possessed great skill in mixing gold with other metals and gave it a natural and perfect appearance that could deceive even the best of silversmiths. The natives were also known for the jewelries made of other precious stones such as carnelian, agate and pearl. Some outstanding examples of Philippine jewelry included necklaces, belts, armlets and rings placed around the waist.

The Americas

Lead mining in the upper Mississippi River region of the U.S., 1865.

There are ancient, prehistoric copper mines along Lake Superior, and metallic copper was still found there, near the surface, in colonial times.   Indegenous peoples availed themselves of this copper starting at least 5,000 years ago," and copper tools, arrowheads, and other artifacts that were part of an extensive native trade network have been discovered. In addition, obsidian, flint, and other minerals were mined, worked, and traded. Early French explorers who encountered the sites made no use of the metals due to the difficulties of transporting them, but the copper was eventually traded throughout the continent along major river routes.

Miners at the Tamarack Mine in Copper Country, Michigan, U.S. in 1905.

In the early colonial history of the Americas, "native gold and silver was quickly expropriated and sent back to Spain in fleets of gold- and silver-laden galleons," the gold and silver originating mostly from mines in Central and South America. Turquoise dated at 700 A.D. was mined in pre-Columbian America; in the Cerillos Mining District in New Mexico, estimates are that "about 15,000 tons of rock had been removed from Mt. Chalchihuitl using stone tools before 1700."

Mining in the United States became prevalent in the 19th century, and the General Mining Act of 1872 was passed to encourage mining of federal lands. As with the California Gold Rush in the mid-19th century, mining for minerals and precious metals, along with ranching, was a driving factor in the Westward Expansion to the Pacific coast. With the exploration of the West, mining camps were established and "expressed a distinctive spirit, an enduring legacy to the new nation;" Gold Rushers would experience the same problems as the Land Rushers of the transient West that preceded them. Aided by railroads, many traveled West for work opportunities in mining. Western cities such as Denver and Sacramento originated as mining towns.

When new areas were explored, it was usually the gold (placer and then load) and then silver that were taken into possession and extracted first. Other metals would often wait for railroads or canals, as coarse gold dust and nuggets do not require smelting and are easy to identify and transport.

Modern Period

In the early 20th century, the gold and silver rush to the western United States also stimulated mining for coal as well as base metals such as copper, lead, and iron. Areas in modern Montana, Utah, Arizona, and later Alaska became predominate suppliers of copper to the world, which was increasingly demanding copper for electrical and households goods. Canada's mining industry grew more slowly than did the United States' due to limitations in transportation, capital, and U.S. competition; Ontario was the major producer of the early 20th century with nickel, copper, and gold.

Meanwhile, Australia experienced the Australian gold rushes and by the 1850s was producing 40% of the world's gold, followed by the establishment of large mines such as the Mount Morgan Mine, which ran for nearly a hundred years, Broken Hill ore deposit (one of the largest zinc-lead ore deposits), and the iron ore mines at Iron Knob. After declines in production, another boom in mining occurred in the 1960s. Now, in the early 21st century, Australia remains a major world mineral producer.

As the 21st century begins, a globalized mining industry of large multinational corporations has arisen. Peak minerals and environmental impacts have also become a concern. Different elements, particularly rare earth minerals, have begun to increase in demand as a result of new technologies.

Mine Development and Lifecycle

The process of mining from discovery of an ore body through extraction of minerals and finally to returning the land to its natural state consists of several distinct steps. The first is discovery of the ore body, which is carried out through prospecting or exploration to find and then define the extent, location and value of the ore body. This leads to a mathematical resource estimation to estimate the size and grade of the deposit.

This estimation is used to conduct a pre-feasibility study to determine the theoretical economics of the ore deposit. This identifies, early on, whether further investment in estimation and engineering studies is warranted and identifies key risks and areas for further work. The next step is to conduct a feasibility study to evaluate the financial viability, the technical and financial risks, and the robustness of the project.

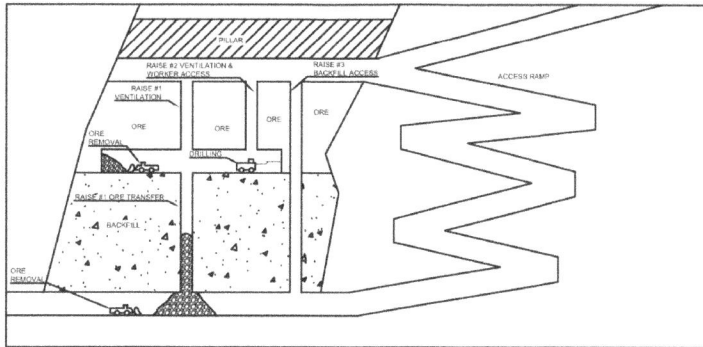

Schematic of a cut and fill mining operation in hard rock.

This is when the mining company makes the decision whether to develop the mine or to walk away from the project. This includes mine planning to evaluate the economically recoverable portion of the deposit, the metallurgy and ore recoverability, marketability and payability of the ore concentrates, engineering concerns, milling and infrastructure costs, finance and equity requirements, and an analysis of the proposed mine from the initial excavation all the way through to reclamation. The proportion of a deposit that is economically recoverable is dependent on the enrichment factor of the ore in the area.

To gain access to the mineral deposit within an area it is often necessary to mine through or remove waste material which is not of immediate interest to the miner. The total movement of ore and waste constitutes the mining process. Often more waste than ore is mined during the life of a mine, depending on the nature and location of the ore body. Waste removal and placement is a major cost to the mining operator, so a detailed characterization of the waste material forms an essential part of the geological exploration program for a mining operation.

Once the analysis determines a given ore body is worth recovering, development begins to create access to the ore body. The mine buildings and processing plants are built, and any necessary equipment is obtained. The operation of the mine to recover the ore begins and continues as long as the company operating the mine finds it economical to do so. Once all the ore that the mine can produce profitably is recovered, reclamation begins to make the land used by the mine suitable for future use.

Mining Techniques

Mining techniques can be divided into two common excavation types: surface mining and sub-surface (underground) mining. Today, surface mining is much more common, and produces, for example, 85% of minerals (excluding petroleum and natural gas) in the United States, including 98% of metallic ores.

Targets are divided into two general categories of materials: *placer deposits*, consisting of valuable minerals contained within river gravels, beach sands, and other unconsolidated materials; and *lode deposits*, where valuable minerals are found in veins, in layers, or in mineral grains generally distributed throughout a mass of actual rock. Both types of ore deposit, placer or lode, are mined by both surface and underground methods.

Underground longwall mining.

Some mining, including much of the rare earth elements and uranium mining, is done by less-common methods, such as in-situ leaching: this technique involves digging neither at the surface nor underground. The extraction of target minerals by this technique requires that they be soluble, e.g., potash, potassium chloride, sodium chloride, sodium sulfate, which dissolve in water. Some minerals, such as copper minerals and uranium oxide, require acid or carbonate solutions to dissolve.

Surface Mining

Surface mining is done by removing (stripping) surface vegetation, dirt, and, if necessary, layers of bedrock in order to reach buried ore deposits. Techniques of surface mining include: open-pit mining, which is the recovery of materials from an open pit in the ground, quarrying, identical to open-pit mining except that it refers to sand, stone and clay; strip mining, which consists of stripping surface layers off to reveal ore/seams underneath; and mountaintop removal, commonly associated with coal mining, which involves taking the top of a mountain off to reach ore deposits at depth. Most (but not all) placer deposits, because of their shallowly buried nature, are mined by surface methods. Finally, landfill mining involves sites where landfills are excavated and processed.

Garzweiler surface mine, Germany

Underground Mining

Sub-surface mining consists of digging tunnels or shafts into the earth to reach buried ore deposits. Ore, for processing, and waste rock, for disposal, are brought to the surface through the tunnels and shafts. Sub-surface mining can be classified by the type of access shafts used, the extraction method or the technique used to reach the mineral deposit. Drift mining utilizes horizontal access tunnels, slope mining uses diagonally sloping access shafts, and shaft mining utilizes vertical access shafts. Mining in hard and soft rock formations require different techniques.

Mantrip used for transporting miners within an underground mine

Other methods include shrinkage stope mining, which is mining upward, creating a sloping underground room, long wall mining, which is grinding a long ore surface underground, and room and pillar mining, which is removing ore from rooms while leaving pillars in place to support the roof of the room. Room and pillar mining often leads to retreat mining, in which supporting pillars are removed as miners retreat, allowing the room to cave in, thereby loosening more ore. Additional sub-surface mining methods include hard rock mining, which is mining of hard rock (igneous, metamorphic or sedimentary) materials, bore hole mining, drift and fill mining, long hole slope mining, sub level caving, and block caving.

Highwall Mining

Caterpillar Highwall Miner HW300 - Technology Bridging Underground and Open Pit Mining

Highwall mining is another form of surface mining that evolved from auger mining. In Highwall mining, the coal seam is penetrated by a continuous miner propelled by a hydraulic Pushbeam Transfer Mechanism (PTM). A typical cycle includes sumping (launch-pushing forward) and shearing (raising and lowering the cutterhead boom to cut the entire height of the coal seam).

As the coal recovery cycle continues, the cutterhead is progressively launched into the coal seam for 19.72 feet (6.01 m). Then, the Pushbeam Transfer Mechanism (PTM) automatically inserts a 19.72-foot (6.01 m) long rectangular Pushbeam (Screw-Conveyor Segment) into the center section of the machine between the Powerhead and the cutterhead. The Pushbeam system can penetrate nearly 1,000 feet (300 m) into the coal seam. One patented Highwall mining systems use augers enclosed inside the Pushbeam that prevent the mined coal from being contaminated by rock debris during the conveyance process. Using a video imaging and/or a gamma ray sensor and/or other Geo-Radar systems like a coal-rock interface detection sensor (CID), the operator can see ahead projection of the seam-rock interface and guide the continuous miner's progress. Highwall mining can produce thousands of tons of coal in contour-strip operations with narrow benches, previously mined areas, trench mine applications and steep-dip seams with controlled water-inflow pump system and/or a gas (inert) venting system.

Machines

The Bagger 288 is a bucket-wheel excavator used in strip mining. It is also the largest land vehicle of all time.

Heavy machinery is used in mining to explore and develop sites, to remove and stockpile over-burden, to break and remove rocks of various hardness and toughness, to process the ore, and to carry out reclamation projects after the mine is closed. Bulldozers, drills, explosives and trucks are all necessary for excavating the land. In the case of placer mining, unconsolidated gravel, or alluvium, is fed into machinery consisting of a hopper and a shaking screen or trommel which frees the desired minerals from the waste gravel. The minerals are then concentrated using sluices or jigs.

A Bucyrus Erie 2570 dragline and CAT 797 haul truck at the North Antelope Rochelle opencut coal mine

Large drills are used to sink shafts, excavate stopes, and obtain samples for analysis. Trams are used to transport miners, minerals and waste. Lifts carry miners into and out of mines, and move rock and ore out, and machinery in and out, of underground mines. Huge trucks, shovels and cranes are employed in surface mining to move large quantities of overburden and ore. Processing plants utilize large crushers, mills, reactors, roasters and other equipment to consolidate the mineral-rich material and extract the desired compounds and metals from the ore.

Processing

Once the mineral is extracted, it is often then processed. The science of extractive metallurgy is a specialized area in the science of metallurgy that studies the extraction of valuable metals from their ores, especially through chemical or mechanical means.

Mineral processing (or mineral dressing) is a specialized area in the science of metallurgy that studies the mechanical means of crushing, grinding, and washing that enable the separation (extractive metallurgy) of valuable metals or minerals from their gangue (waste material). Processing of placer ore material consists of gravity-dependent methods of separation, such as sluice boxes. Only minor shaking or washing may be necessary to disaggregate (unclump) the sands or gravels before processing. Processing of ore from a lode mine, whether it is a surface or subsurface mine, requires that the rock ore be crushed and pulverized before extraction of the valuable minerals begins. After lode ore is crushed, recovery of the valuable minerals is done by one, or a combination of several, mechanical and chemical techniques.

Since most metals are present in ores as oxides or sulfides, the metal needs to be reduced to its metallic form. This can be accomplished through chemical means such as smelting or through electrolytic reduction, as in the case of aluminium. Geometallurgy combines the geologic sciences with extractive metallurgy and mining.

Environmental Effects

Iron hydroxide precipitate stains a stream receiving acid drainage from surface coal mining.

Environmental issues can include erosion, formation of sinkholes, loss of biodiversity, and contamination of soil, groundwater and surface water by chemicals from mining processes. In some cases, additional forest logging is done in the vicinity of mines to create space for the storage of the created debris and soil. Contamination resulting from leakage of chemicals can also affect the health of the local population if not properly controlled. Extreme examples of pollution from mining activities include coal fires, which can last for years or even decades, producing massive amounts of environmental damage.

Mining companies in most countries are required to follow stringent environmental and rehabilitation codes in order to minimize environmental impact and avoid impacting human health. These codes and regulations all require the common steps of environmental impact assessment, development of environmental management plans, mine closure planning (which must be done before the start of mining operations), and environmental monitoring during operation and after closure. However, in some areas, particularly in the developing world, government regulations may not be well enforced.

For major mining companies and any company seeking international financing, there are a number of other mechanisms to enforce good environmental standards. These generally relate to financing standards such as the Equator Principles, IFC environmental standards, and criteria for Socially responsible investing. Mining companies have used this oversight from the financial sector to argue for some level of industry self-regulation. In 1992, a Draft Code of Conduct for Transnational Corporations was proposed at the Rio Earth Summit by the UN Centre for Transnational Corporations (UNCTC), but the Business Council for Sustainable Development (BCSD) together with the International Chamber of Commerce (ICC) argued successfully for self-regulation instead.

This was followed by the Global Mining Initiative which was begun by nine of the largest metals and mining companies and which led to the formation of the International Council on Mining and Metals, whose purpose was to "act as a catalyst" in an effort to improve social and environmental performance in the mining and metals industry internationally. The mining industry has provided funding to various conservation groups, some of which have been working with conservation agendas that are at odds with an emerging acceptance of the rights of indigenous people – particularly the right to make land-use decisions.

Certification of mines with good practices occurs through the International Organization for Standardization (ISO). For example, ISO 9000 and ISO 14001, which certify an "auditable environmental management system", involve short inspections, although they have been accused of lacking rigor. Certification is also available through Ceres' Global Reporting Initiative, but these reports are voluntary and unverified. Miscellaneous other certification programs exist for various projects, typically through nonprofit groups.

The purpose of a 2012 EPS PEAKS paper was to provide evidence on policies managing ecological costs and maximise socio-economic benefits of mining using host country regulatory initiatives. It found existing literature suggesting donors encourage developing countries to:

- Make the environment-poverty link and introduce cutting-edge wealth measures and natural capital accounts.

- Reform old taxes in line with more recent financial innovation, engage directly with the companies, enacting land use and impact assessments, and incorporate specialised support and standards agencies.

- Set in play transparency and community participation initiatives using the wealth accrued.

Waste

Ore mills generate large amounts of waste, called tailings. For example, 99 tons of waste are generated per ton of copper, with even higher ratios in gold mining - because only 5.3 g of gold is extracted per ton of ore, a ton of gold produces 200,000 tons of tailings. These tailings can be toxic. Tailings, which are usually produced as a slurry, are most commonly dumped into ponds made from naturally existing valleys. These ponds are secured by impoundments (dams or embankment dams). In 2000 it was estimated that 3,500 tailings impoundments existed, and that every year, 2 to 5 major failures and 35 minor failures occurred; for example, in the Marcopper mining disaster at least 2 million tons of tailings were released into a local river. Subaqueous tailings disposal is another option. The mining industry has argued that submarine tailings disposal (STD), which disposes of tailings in the sea, is ideal because it avoids the risks of tailings ponds; although the practice is illegal in the United States and Canada, it is used in the developing world.

The waste is classified as either sterile or mineralised, with acid generating potential, and the movement and storage of this material forms a major part of the mine planning process. When the mineralised package is determined by an economic cut-off, the near-grade mineralised waste is usually dumped separately with view to later treatment should market conditions change and it becomes economically viable. Civil engineering design parameters are used in the design of the waste dumps, and special conditions apply to high-rainfall areas and to seismically active areas. Waste dump designs must meet all regulatory requirements of the country in whose jurisdiction the mine is located. It is also common practice to rehabilitate dumps to an internationally acceptable standard, which in some cases means that higher standards than the local regulatory standard are applied.

Renewable Energy and Mining

Many mining sites are remote and not connected to the grid. Electricity is typically generated with diesel generators. Due to high transportation cost and theft during transportation the cost for generating electricity is normally high. Renewable energy applications are becoming an alternative or amendment. Both solar and wind power plants can contribute in saving diesel costs at mining sites. Renewable energy applications have been built at mining sites. Cost savings can reach up to 70%.

Mining Industry

Mining exists in many countries. London is known as the capital of global "mining houses" such as Rio Tinto Group, BHP Billiton, and Anglo American PLC. The US mining industry is also large, but it is dominated by the coal and other nonmetal minerals (e.g., rock and sand), and various regulations have worked to reduce the significance of mining in the United States. In 2007 the total market capitalization of mining companies was reported at US$962 billion, which compares to a total global market cap of publicly traded companies of about US$50 trillion in 2007. In 2002, Chile and Peru were reportedly the major mining countries of South America. The mineral industry of

Africa includes the mining of various minerals; it produces relatively little of the industrial metals copper, lead, and zinc, but according to one estimate has as a percent of world reserves 40% of gold, 60% of cobalt, and 90% of the world's platinum group metals. Mining in India is a significant part of that country's economy. In the developed world, mining in Australia, with BHP Billiton founded and headquartered in the country, and mining in Canada are particularly significant. For rare earth minerals mining, China reportedly controlled 95% of production in 2013.

The Bingham Canyon Mine of Rio Tinto's subsidiary, Kennecott Utah Copper.

While exploration and mining can be conducted by individual entrepreneurs or small businesses, most modern-day mines are large enterprises requiring large amounts of capital to establish. Consequently, the mining sector of the industry is dominated by large, often multinational, companies, most of them publicly listed. It can be argued that what is referred to as the 'mining industry' is actually two sectors, one specializing in exploration for new resources and the other in mining those resources. The exploration sector is typically made up of individuals and small mineral resource companies, called "juniors", which are dependent on venture capital. The mining sector is made up of large multinational companies that are sustained by production from their mining operations. Various other industries such as equipment manufacture, environmental testing, and metallurgy analysis rely on, and support, the mining industry throughout the world. Canadian stock exchanges have a particular focus on mining companies, particularly junior exploration companies through Toronto's TSX Venture Exchange; Canadian companies raise capital on these exchanges and then invest the money in exploration globally. Some have argued that below juniors there exists a substantial sector of illegitimate companies primarily focused on manipulating stock prices.

Mining operations can be grouped into five major categories in terms of their respective resources. These are oil and gas extraction, coal mining, metal ore mining, nonmetallic mineral mining and quarrying, and mining support activities. Of all of these categories, oil and gas extraction remains one of the largest in terms of its global economic importance. Prospecting potential mining sites, a vital area of concern for the mining industry, is now done using sophisticated new technologies such as seismic prospecting and remote-sensing satellites. Mining is heavily affected by the prices of the commodity minerals, which are often volatile. The 2000s commodities boom ("commodities supercycle") increased the prices of commodities, driving aggressive mining. In addition, the price of gold increased dramatically in the 2000s, which increased gold mining; for example, one study found that conversion of forest in the Amazon increased six-fold from the period 2003–2006 (292 ha/yr) to the period 2006–2009 (1,915 ha/yr), largely due to artisanal mining.

Corporate Classifications

Mining companies can be classified based on their size and financial capabilities:

- Major companies are considered to have an adjusted annual mining-related revenue of more than US$500 million, with the financial capability to develop a major mine on its own.

- Intermediate companies have at least $50 million in annual revenue but less than $500 million.

- Junior companies rely on equity financing as their principal means of funding exploration. Juniors are mainly pure exploration companies, but may also produce minimally, and do not have a revenue exceeding US$50 million.

Regulation and Governance

New regulation and process of legislative reforms aims to enrich the harmonization and stability of the mining sector in mineral-rich countries. The new legislation for mining industry in the African countries still appears as an emerging issue with a potential to be solved, until a consensus is reached on the best approach. By the beginning of the 21st century the booming and more complex mining sector in mineral-rich countries provided only slight benefits to local communities in terms of sustainability. Increasing debates and influence by NGOs and communities appealed for a new program which would have had also included a disadvantaged communities, and would have had worked towards sustainable development even after mine closure (included transparency and revenue management). By the early 2000s, community development issues and resettlements became mainstreamed in World Bank mining projects. Mining-industry expansion after an increase of mineral prices in 2003 and also potential fiscal revenues in those countries created an omission in the other economic sectors in terms of finances and development. Furthermore, it had highlighted regional and local demand of mining-revenues and lack of ability of sub-national governments to use the revenues. The Fraser Institute (a Canadian think tank) has highlighted the environmental protection laws in developing countries, as well as the voluntary efforts by mining companies to improve their environmental impact.

In 2007 the Extractive Industries Transparency Initiative (EITI) was mainstreamed in all countries cooperating with the World Bank in mining industry reform. The EITI is operating and implementing with a support of EITI Multi-Donor Trust Fund, managed by The World Bank. The Extractive Industries Transparency Initiative (EITI) aims to increase transparency in transactions between governments and companies within extractive industries by monitoring the revenues and benefits between industries and recipient governments. The entrance process is voluntary for each country and is being monitored by multi-stakeholders involving government, private companies and civil society representatives, responsible for disclosure and dissemination of the reconciliation report; however, the competitive disadvantage of company-by company public report is for some of the businesses in Ghana, the main constraint. Therefore, the outcome assessment in terms of failure or success of the new EITI regulation does not only "rest on the government's shoulders" but also on civil society and companies.

On the other hand, criticism points out two main implementation issues; inclusion or exclusion of artisanal mining and small-scale mining (ASM) from the EITI and how to deal with "non-cash" payments made by companies to subnational governments. Furthermore, disproportion of the revenues mining industry creates to the comparatively small number of people that it employs, causes another controversy. The issue of artisanal mining is clearly an issue in EITI Countries such as the Central African Republic, D.R. Congo, Guinea, Liberia and Sierra Leone – i.e. almost half of the mining countries implementing the EITI. Among other things, limited scope of the EITI involving disparity in terms of knowledge of the industry and negotiation skills, thus far flexibility of the policy (e.g. liberty of the countries to expand beyond the minimum requirements and adapt it to their needs), creates another risk of unsuccessful implementation. Public awareness increase, where government should act as a bridge between public and initiative for a successful outcome of the policy is an important element to be considered.

World Bank

The World Bank has been involved in mining since 1955, mainly through grants from its International Bank for Reconstruction and Development, with the Bank's Multilateral Investment Guarantee Agency offering political risk insurance. Between 1955 and 1990 it provided about $2 billion to fifty mining projects, broadly categorized as reform and rehabilitation, greenfield mine construction, mineral processing, technical assistance, and engineering. These projects have been criticized, particularly the Ferro Carajas project of Brazil, begun in 1981. The World Bank established mining codes intended to increase foreign investment; in 1988 it solicited feedback from 45 mining companies on how to increase their involvement.

In 1992 the World Bank began to push for privatization of government-owned mining companies with a new set of codes, beginning with its report *The Strategy for African Mining*. In 1997, Latin America's largest miner Companhia Vale do Rio Doce (CVRD) was privatized. These and other developments such as the Philippines 1995 Mining Act led the bank to publish a third report (*Assistance for Minerals Sector Development and Reform in Member Countries*) which endorsed mandatory environment impact assessments and attention to the concerns of the local population. The codes based on this report are influential in the legislation of developing nations. The new codes are intended to encourage development through tax holidays, zero custom duties, reduced income taxes, and related measures. The results of these codes were analyzed by a group from the University of Quebec, which concluded that the codes promote foreign investment but "fall very short of permitting sustainable development". The observed negative correlation between natural resources and economic development is known as the resource curse.

Safety

Safety has long been a concern in the mining business, especially in sub-surface mining. The Courrières mine disaster, Europe's worst mining accident, involved the death of 1,099 miners in Northern France on March 10, 1906. This disaster was surpassed only by the Benxihu Colliery accident in China on April 26, 1942, which killed 1,549 miners. While mining today is substantially safer than it was in previous decades, mining accidents still occur. Government figures indicate that

5,000 Chinese miners die in accidents each year, while other reports have suggested a figure as high as 20,000. Mining accidents continue worldwide, including accidents causing dozens of fatalities at a time such as the 2007 Ulyanovskaya Mine disaster in Russia, the 2009 Heilongjiang mine explosion in China, and the 2010 Upper Big Branch Mine disaster in the United States.

There are numerous occupational hazards associated with mining, including exposure to rockdust which can lead to diseases such as silicosis, asbestosis, and pneumoconiosis. Gases in the mine can lead to asphyxiation and could also be ignited. Mining equipment can generate considerable noise, putting workers at risk for hearing loss. Cave-ins, rock falls, and exposure to excess heat are also known hazards.

Proper ventilation, hearing protection, and spraying equipment with water are important safety practices in mines.

Records

Chuquicamata, Chile, site of the largest circumference and second deepest open pit copper mine in the world.

As of 2008, the deepest mine in the world is TauTona in Carletonville, South Africa at 3.9 kilometres (2.4 mi), replacing the neighboring Savuka Mine in the North West Province of South Africa at 3,774 metres (12,382 ft). East Rand Mine in Boksburg, South Africa briefly held the record at 3,585 metres (11,762 ft), and the first mine declared the deepest in the world was also TauTona when it was at 3,581 metres (11,749 ft).

The Moab Khutsong gold mine in North West Province (South Africa) has the world's longest winding steel wire rope, able to lower workers to 3,054 metres (10,020 ft) in one uninterrupted four-minute journey.

The deepest mine in Europe is the 16th shaft of the uranium mines in Příbram, Czech Republic at 1,838 metres (6,030 ft), second is Bergwerk Saar in Saarland, Germany at 1,750 metres (5,740 ft).

The deepest open-pit mine in the world is Bingham Canyon Mine in Bingham Canyon, Utah, United States at over 1,200 metres (3,900 ft). The largest and second deepest open-pit copper mine in the world is Chuquicamata in Chuquicamata, Chile at 900 metres (3,000 ft), 443,000 tons of copper and 20,000 tons of molybdenum produced annually.

The deepest open-pit mine with respect to sea level is Tagebau Hambach in Germany, where the base of the pit is 293 metres (961 ft) below sea level.

The largest underground mine is Kiirunavaara Mine in Kiruna, Sweden. With 450 kilometres (280 mi) of roads, 40 million tonnes of ore produced yearly, and a depth of 1,270 metres (4,170 ft), it is also one of the most modern underground mines. The deepest borehole in the world is Kola Superdeep Borehole at 12,262 metres (40,230 ft). This, however, is not a matter of mining but rather related to scientific drilling.

Metal Reserves and Recycling

During the twentieth century, the variety of metals used in society grew rapidly. Today, the development of major nations such as China and India and advances in technologies are fueling an ever greater demand. The result is that metal mining activities are expanding and more and more of the world's metal stocks are above ground in use rather than below ground as unused reserves. An example is the in-use stock of copper. Between 1932 and 1999, copper in use in the USA rose from 73 kilograms (161 lb) to 238 kilograms (525 lb) per person.

95% of the energy used to make aluminium from bauxite ore is saved by using recycled material. However, levels of metals recycling are generally low. In 2010, the International Resource Panel, hosted by the United Nations Environment Programme (UNEP), published reports on metal stocks that exist within society and their recycling rates.

The report's authors observed that the metal stocks in society can serve as huge mines above ground. However, they warned that the recycling rates of some rare metals used in applications such as mobile phones, battery packs for hybrid cars, and fuel cells are so low that unless future end-of-life recycling rates are dramatically stepped up these critical metals will become unavailable for use in modern technology.

As recycling rates are low and so much metal has already been extracted, some landfills now contain higher concentrations of metal than mines themselves. This is especially true with aluminium, found in cans, and precious metals in discarded electronics. Furthermore, waste after 15 years has still not broken down, so less processing would be required when compared to mining ores. A study undertaken by Cranfield University has found £360 million of metals could be mined from just 4 landfill sites. There is also up to 20MW/kg of energy in waste, potentially making the re-extraction more profitable. However, although the first landfill mine opened in Tel Aviv, Israel in 1953, little work has followed due to the abundance of accessible ores.

References

- "Chuquicamata's Life Underground Will Cost a Fortune, but is Likely to Pay Off for Codelco | Copper Investing News". 2015-04-06. Archived from the original on April 6, 2015. Retrieved 2015-06-11.

- "Study shows around £360 million of metals could be mined from just four landfill sites". www.rebnews.com. Retrieved 2015-06-11.

- "Assessing the opportunities of landfill mining - Research database - University of Groningen". www.rug.nl. Retrieved 2015-06-11.

- "Mining, Quarrying & Prospecting: The Difference between Mining, Quarrying & Prospecting". mqp-geotek. blogspot.co.uk. Retrieved 2015-06-11.

- "http://www.macfarlanes.com/media/1467/landfill-mining-new-opportunities-ahead.pdf" (PDF). www.mac-farlanes.com. Retrieved 2015-06-11.

- "The Independent, 20 Jan. 2007: ''The end of a Celtic tradition: the last gold miner in Wales''". News.independent.co.uk. 2007-01-20. Archived from the original on July 6, 2008. Retrieved 2013-06-22.

- Cambell, Bonnie (2008). "Regulation & Legitimacy in the Mining Industry in Africa: Where does" (PDF). Review of African Political Economy. 35 (3): 367–389. doi:10.1080/03056240802410984. Retrieved 7 April 2011.

- The World Bank. ces.worldbank.org/INTOGMC/Resources/336099-1288881181404/7530465-128888120-7444/eifd19_mining_sector_reform.pdf "The World Bank's Evolutionary Approach to Mining Sector Reform" Check |url= value (help) (PDF). Retrieved 4 April 2011.

- Business and Human Right Resource Centre (2009). "Principles: Extractive Industries Transparency Initiative (EITI)". Retrieved 6 April 2011.

- World Bank's Oil, Gas and Mining Policy and Operations Unit (COCPO). "Advancing the EITI in the Mining Sector: Implementation Issues" (PDF). Retrieved 6 April 2011.

- Revenue Watch Institute 2010. "Promoting Transparency in the Extractive Sectors: An EITI Training for Tanzania Legislators". Archived from the original on July 20, 2011. Retrieved 6 April 2011.

- Chapin, Mac (2004-10-15). "A Challenge to Conservationists: Can we protect natural habitats without abusing the people who live in them?". World Watch Magazine. 6. 17. Retrieved 2010-02-18.

Types of Mining

Mining is practiced in many countries and can be grouped into major categories such as coal mining, gold mining, metal ore mining and placer mining. Mining is best understood in confluence with the major topics listed in the following chapter. The topics discussed in the chapter are of great importance to broaden the existing knowledge on mining.

Coal Mining

The goal of coal mining is to obtain coal from the ground. Coal is valued for its energy content, and, since the 1880s, has been widely used to generate electricity. Steel and cement industries use coal as a fuel for extraction of iron from iron ore and for cement production. In the United Kingdom, and South Africa, a coal mine and its structures are a colliery. In Australia, "colliery" generally refers to an underground coal mine. In the United States "colliery" has historically been used to describe a coal mine operation, but the word today is not commonly used.

A painting depicting men leaving a UK colliery at the close of a shift.

Coal mining has had many developments over the recent years, from the early days of men tunneling, digging and manually extracting the coal on carts, to large open cut and long wall mines. Mining at this scale requires the use of draglines, trucks, conveyors, jacks and shearers.

History

Small scale mining of surface deposits dates back thousands of years. For example, in Roman Britain, the Romans were exploiting all major coalfields (save those of North and South Staffordshire) by the late 2nd century AD. While much of its use remained local, a lively trade developed along the North Sea coast supplying coal to Yorkshire and London.

Ships were used to haul coal.

The Industrial Revolution, which began in Britain in the 18th century, and later spread to continental Europe and North America, was based on the availability of coal to power steam engines. International trade expanded exponentially when coal-fed steam engines were built for the railways and steamships. The new mines that grew up in the 19th century depended on men and children to work long hours in often dangerous working conditions. There were many coalfields, but the oldest were in Newcastle and Durham, South Wales, the Central Belt of Scotland and the Midlands, such as those at Coalbrookdale.

Children would usually work as trappers; this is where they had to open and close trap doors to allow mine carts in. This made sure no harmful gases built up in the mine.

The oldest continuously worked deep-mine in the United Kingdom is Tower Colliery in South Wales valleys in the heart of the South Wales coalfield. This colliery was developed in 1805, and its miners bought it out at the end of the 20th century, to prevent it from being closed. Tower Colliery was finally closed on 25 January 2008, although production continues at the Aberpergwm drift mine owned by Walter Energy.

Coal was mined in America in the early 18th century, and commercial mining started around 1730 in Midlothian, Virginia.

Coal-cutting machines were invented in the 1880s. Before this invention, coal was mined from underground with a pick and shovel. By 1912, surface mining was conducted with steam shovels designed for coal mining.

Methods of Extraction

The most economical method of coal extraction from coal seams depends on the depth and quality of the seams, and the geology and environmental factors. Coal mining processes are differentiated by whether they operate on the surface or underground. Many coals extracted from both surface and underground mines require washing in a coal preparation plant. Technical and economic feasibility are evaluated based on the following: regional geological conditions; overburden characteristics; coal seam continuity, thickness, structure, quality, and depth; strength of materials above and below the seam for roof and floor conditions; topography (especially altitude and slope); climate; land ownership as it affects the availability of land for mining and access; surface drainage patterns; ground water conditions; availability of labor and materials; coal purchaser requirements in terms of tonnage, quality, and destination; and capital investment requirements.

Surface mining and deep underground mining are the two basic methods of mining. The choice of mining method depends primarily on depth of burial, density of the overburden and thickness of the coal seam. Seams relatively close to the surface, at depths less than approximately 180 ft (50 m), are usually surface mined.

Coal that occurs at depths of 180 to 300 ft (50 to 100 m) are usually deep mined, but in some cases surface mining techniques can be used. For example, some western U.S. coal that occur at depths in excess of 200 ft (60 m) are mined by the open pit methods, due to thickness of the seam 60–90 feet (20–30 m). Coals occurring below 300 ft (100 m) are usually deep mined. However, there are open pit mining operations working on coal seams up to 1000–1500 feet (300–450 m) below ground level, for instance Tagebau Hambach in Germany.

Modern Surface Mining

Trucks loaded with coal at the Cerrejón coal mine in Colombia

When coal seams are near the surface, it may be economical to extract the coal using open cut (also referred to as open cast, open pit, or strip) mining methods. Open cast coal mining recovers a greater proportion of the coal deposit than underground methods, as more of the coal seams in the strata may be exploited. Large open cast mines can cover an area of many square kilometers and use very large pieces of equipment. This equipment can include the following: Draglines which operate by removing the overburden, power shovels, large trucks in which transport overburden and coal, bucket wheel excavators, and conveyors. In this mining method, explosives are first used in order to break through the surface, or overburden, of the mining area. The overburden is then removed by draglines or by shovel and truck. Once the coal seam is exposed, it is drilled, fractured and thoroughly mined in strips. The coal is then loaded on to large trucks or conveyors for transport to either the coal preparation plant or directly to where it will be used.

Most open cast mines in the United States extract bituminous coal. In Canada (BC), Australia and South Africa, open cast mining is used for both thermal and metallurgical coals. In New South Wales open casting for steam coal and anthracite is practised. Surface mining accounts for around 80 percent of production in Australia, while in the US it is used for about 67 percent of production. Globally, about 40 percent of coal production involves surface mining.

Strip Mining

Strip mining exposes coal by removing earth above each coal seam. This earth is referred to as overburden and is removed in long strips. The overburden from the first strip is deposited in an area outside the planned mining area and referred to as out-of-pit dumping. Overburden from subsequent strips are deposited in the void left from mining the coal and overburden from the previous strip. This is referred to as in-pit dumping.

It is often necessary to fragment the overburden by use of explosives. This is accomplished by drilling holes into the overburden, filling the holes with explosives, and detonating the explosive. The overburden is then removed, using large earth-moving equipment, such as draglines, shovel and trucks, excavator and trucks, or bucket-wheels and conveyors. This overburden is put into the previously mined (and now empty) strip. When all the overburden is removed, the underlying coal seam will be exposed (a 'block' of coal). This block of coal may be drilled and blasted (if hard) or otherwise loaded onto trucks or conveyors for transport to the coal preparation (or wash) plant. Once this strip is empty of coal, the process is repeated with a new strip being created next to it. This method is most suitable for areas with flat terrain.

Equipment to be used depends on geological conditions. For example, to remove overburden that is loose or unconsolidated, a bucket wheel excavator might be the most productive. The life of some area mines may be more than 50 years.

Contour Mining

The contour mining method consists of removing overburden from the seam in a pattern following the contours along a ridge or around the hillside. This method is most commonly used in areas with rolling to steep terrain. It was once common to deposit the spoil on the downslope side of the bench thus created, but this method of spoil disposal consumed much additional land and created severe landslide and erosion problems. To alleviate these problems, a variety of methods were devised to use freshly cut overburden to refill mined-out areas. These haul-back or lateral movement methods generally consist of an initial cut with the spoil deposited downslope or at some other site and spoil from the second cut refilling the first. A ridge of undisturbed natural material 15 to 20 ft (5–6 m) wide is often intentionally left at the outer edge of the mined area. This barrier adds stability to the reclaimed slope by preventing spoil from slumping or sliding downhill.

The limitations of contour strip mining are both economic and technical. When the operation reaches a predetermined stripping ratio (tons of overburden/tons of coal), it is not profitable to continue. Depending on the equipment available, it may not be technically feasible to exceed a certain height of highwall. At this point, it is possible to produce more coal with the augering method in which spiral drills bore tunnels into a highwall laterally from the bench to extract coal without removing the overburden.

Mountaintop Removal Mining

Mountaintop coal mining is a surface mining practice involving removal of mountaintops to expose coal seams, and disposing of associated mining overburden in adjacent "valley fills." Valley fills occur in steep terrain where there are limited disposal alternatives.

Mountaintop removal combines area and contour strip mining methods. In areas with rolling or steep terrain with a coal seam occurring near the top of a ridge or hill, the entire top is removed in a series of parallel cuts. Overburden is deposited in nearby valleys and hollows. This method usually leaves ridge and hill tops as flattened plateaus. The process is highly controversial for the drastic changes in topography, the practice of creating *head-of-hollow-fills*, or filling in valleys with mining debris, and for covering streams and disrupting ecosystems.

Spoil is placed at the head of a narrow, steep-sided valley or hollow. In preparation for filling this area, vegetation and soil are removed and a rock drain constructed down the middle of the area to be filled, where a natural drainage course previously existed. When the fill is completed, this underdrain will form a continuous water runoff system from the upper end of the valley to the lower end of the fill. Typical head-of-hollow fills are graded and terraced to create permanently stable slopes.

Underground Mining

Coal wash plant in Clay County, Kentucky

Most coal seams are too deep underground for opencast mining and require underground mining, a method that currently accounts for about 60 percent of world coal production. In deep mining, the room and pillar or bord and pillar method progresses along the seam, while pillars and timber are left standing to support the mine roof. Once room and pillar mines have been developed to a stopping point (limited by geology, ventilation, or economics), a supplementary version of room and pillar mining, termed second mining or retreat mining, is commonly started. Miners remove the coal in the pillars, thereby recovering as much coal from the coal seam as possible. A work area involved in pillar extraction is called a pillar section.

Modern pillar sections use remote-controlled equipment, including large hydraulic mobile roof-supports, which can prevent cave-ins until the miners and their equipment have left a work area. The mobile roof supports are similar to a large dining-room table, but with hydraulic jacks for legs. After the large pillars of coal have been mined away, the mobile roof support's legs shorten and it is withdrawn to a safe area. The mine roof typically collapses once the mobile roof supports leave an area.

Remote Joy HM21 Continuous Miner used underground

There are six principal methods of underground mining:

- Longwall mining accounts for about 50 percent of underground production. The longwall shearer has a face of 1,000 feet (300 m) or more. It is a sophisticated machine with a rotating drum that moves mechanically back and forth across a wide coal seam. The loosened coal falls on to an armored chain conveyor or pan line that takes the coal to the conveyor belt for removal from the work area. Longwall systems have their own hydraulic roof supports which advance with the machine as mining progresses. As the longwall mining equipment moves forward, overlying rock that is no longer supported by coal is allowed to fall behind the operation in a controlled manner. The supports make possible high levels of production and safety. Sensors detect how much coal remains in the seam while robotic controls enhance efficiency. Longwall systems allow a 60-to-100 percent coal recovery rate when surrounding geology allows their use. Once the coal is removed, usually 75 percent of the section, the roof is allowed to collapse in a safe manner.

- Continuous mining utilizes a Continuous Miner Machine with a large rotating steel drum equipped with tungsten carbide picks that scrape coal from the seam. Operating in a "room and pillar" (also known as "bord and pillar") system—where the mine is divided into a series of 20-to-30 foot (5–10 m) "rooms" or work areas cut into the coalbed—it can mine as much as 14 tons of coal a minute, more than a non-mechanised mine of the 1920s would produce in an entire day. Continuous miners account for about 45 percent of underground coal production. Conveyors transport the removed coal from the seam. Remote-controlled continuous miners are used to work in a variety of difficult seams and conditions, and robotic versions controlled by computers are becoming increasingly common. Continuous mining is a misnomer, as room and pillar coal mining is very cyclical. In the US, one can generally cut 20 ft or 6 meters (or a bit more with MSHA permission) (12 meters or roughly 40 ft in South Africa before the Continuous Miner goes out and the roof is supported by the Roof Bolter), after which, the face has to be serviced, before it can be advanced again. During servicing, the "continuous" miner moves to another face. Some continuous miners can bolt and rock dust the face (two major components of servicing) while cutting coal, while a trained crew may be able to advance ventilation, to truly earn the "continuous" label. However, very few mines are able to achieve it. Most continuous mining machines in use in the US lack the ability to bolt and dust. This may partly be because incorporation of bolting makes the machines wider, and therefore, less maneuverable.The goal of coal mining is to obtain coal from the ground. Coal is valued for its energy content, and, since the 1880s, has been widely used to generate electricity. Steel and cement industries use coal as a fuel for extraction of iron from iron ore and for cement production. In the United Kingdom, and South Africa, a coal mine and its struct.

- Room and pillar mining consists of coal deposits that are mined by cutting a network of rooms into the coal seam. Pillars of coal are left behind in order to keep up the roof. The pillars can make up to forty percent of the total coal in the seam, however where there was space to leave head and floor coal there is evidence from recent open cast excavations that 18th century operators used a variety of room and pillar techniques to remove 92 percent of the *in situ* coal. However, this can be extracted at a later stage.

- Blast mining or conventional mining, is an older practice that uses explosives such as dynamite to break up the coal seam, after which the coal is gathered and loaded on to shuttle cars or conveyors for removal to a central loading area. This process consists of a series of operations that begins with "cutting" the coalbed so it will break easily when blasted with explosives. This type of mining accounts for less than 5 percent of total underground production in the US today.

- Shortwall mining, a method currently accounting for less than 1 percent of deep coal production, involves the use of a continuous mining machine with movable roof supports, similar to longwall. The continuous miner shears coal panels 150 to 200 feet (40 to 60 m) wide and more than a half-mile (1 km) long, having regard to factors such as geological strata.

- Retreat mining is a method in which the pillars or coal ribs used to hold up the mine roof are extracted; allowing the mine roof to collapse as the mining works back towards the entrance. This is one of the most dangerous forms of mining, owing to imperfect predictability of when the roof will collapse and possibly crush or trap workers in the mine.

Production

Coal is mined commercially in over 50 countries. Over 7,036 Mt/yr of hard coal is currently produced, a substantial increase over the past 25 years. In 2006, the world production of brown coal and lignite was slightly over 1,000 Mt, with Germany the world's largest brown coal producer at 194.4 Mt, and China second at 100.6 Mt.

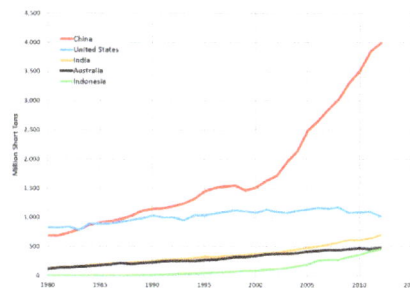

Coal production trends 1980-2012 in the top five coal-producing countries (US EIA)

Coal mine in China

Coal mine in Australia

Coal production has grown fastest in Asia, while Europe has declined. The top coal mining nations (figures in brackets are 2009 estimate of total coal production in millions of tons) are:

- China (3,050 Mt)

- United States (973 Mt)

- India (557 Mt)

- Australia (409 Mt)

- Russia (298 Mt)

- Indonesia (252 Mt)

- South Africa (250 Mt)

- Poland (135 Mt)

- Kazakhstan (101 Mt)

- Colombia (75 Mt)

Most coal production is used in the country of origin, with around 16 percent of hard coal production being exported.

Global coal production is expected to reach 7,000 Mt/yr in 2030 (Update required, world coal production is already past 7,000 Mt/yr and by 2030 will probably be closer to 13,000 Mt/yr), with China accounting for most of this increase. Steam coal production is projected to reach around 5,200 Mt/yr; coking coal 620 Mt/yr; and brown coal 1,200 Mt/yr.

Coal reserves are available in almost every country worldwide, with recoverable reserves in around 70 countries. At current production levels, proven coal reserves are estimated to last 147 years. However, production levels are by no means level, and are in fact increasing and some estimates are that peak coal could arrive in many countries such as China and America by around 2030. Coal reserves are usually stated as either (1) "Resources" ("measured" + "indicated" + "inferred" = "resources", and then, a smaller number, often only 10-20% of "resources," (2) "Run of Mine" (ROM) reserves, and finally (3) "marketable reserves", which may be only 60% of ROM reserves. The standards for reserves are set by stock exchanges, in consultation with industry associations. For example, in ASEAN countries reserves standards follow the Australasian Joint Ore Reserves Committee Code (JORC) used by the Australian Securities Exchange.

Modern Mining

Laser profiling of a minesite by a coal miner using a Maptek I-site laser scanner in 2014

Technological advancements have made coal mining today more productive than it has ever been. To keep up with technology and to extract coal as efficiently as possible modern mining personnel must be highly skilled and well trained in the use of complex, state-of-the-art instruments and equipment. Many jobs require four-year university degrees. Computer knowledge has also become greatly valued within the industry as most of the machines and safety monitors are computerized.

The use of sophisticated sensing equipment to monitor air quality is common and has replaced the use of small animals such as canaries, often referred to as "miner's canaries".

In the United States, the increase in technology has significantly decreased the mining workforce from 335,000 coal miners working at 7,200 mines fifty years ago to 104,824 miners working in fewer than 2,000 mines today.

Safety

Dangers to Miners

The Farmington coal mine disaster kills 78. West Virginia, US, 1968.

Historically, coal mining has been a very dangerous activity and the list of historical coal mining disasters is a long one. In the US alone, more than 100,000 coal miners were killed in accidents in the twentieth century, 90 percent of the fatalities occurring in the first half of the century. More than 3,200 died in 1907 alone.

Open cut hazards are principally mine wall failures and vehicle collisions; underground mining hazards include suffocation, gas poisoning, roof collapse, rock burst, outbursts, and gas explosions.

Firedamp explosions can trigger the much-more-dangerous coal dust explosions, which can engulf an entire pit. Most of these risks can be greatly reduced in modern mines, and multiple fatality incidents are now rare in some parts of the developed world. Modern mining in the US results in approximately 30 deaths per year due to mine accidents.

However, in lesser developed countries and some developing countries, many miners continue to die annually, either through direct accidents in coal mines or through adverse health consequences from working under poor conditions. China, in particular, has the highest number of coal mining related deaths in the world, with official statistics claiming that 6,027 deaths occurred in 2004. To compare, 28 deaths were reported in the US in the same year. Coal production in China is twice that in the US, while the number of coal miners is around 50 times that of the US, making deaths in coal mines in China 4 times as common per worker (108 times as common per unit output) as in the US.

Mine disasters have still occurred in recent years in the US, Examples include the Sago Mine disaster of 2006, and the 2007 mine accident in Utah's Crandall Canyon Mine, where nine miners were killed and six entombed. In the decade 2005-2014, US coal mining fatalities averaged 28 per year.The most fatalities during the 2005-2014 decade were 48 in 2010, the year of the Upper Big Branch Mine disaster in West Virginia, which killed 29 miners.

Miners can be regularly monitored for reduced lung function due to coal dust exposure using spirometry.

Chronic lung diseases, such as pneumoconiosis (black lung) were once common in miners, leading to reduced life expectancy. In some mining countries black lung is still common, with 4,000 new cases of black lung every year in the US (4 percent of workers annually) and 10,000 new cases every year in China (0.2 percent of workers). Rates may be higher than reported in some regions.

Build-ups of a hazardous gas are known as damps, possibly from the German word "Dampf" which means steam or vapor:

- Black damp: a mixture of carbon dioxide and nitrogen in a mine can cause suffocation, and is formed as a result of corrosion in enclosed spaces so removing oxygen from the atmosphere.

- After damp: similar to black damp, after damp consists of carbon monoxide, carbon dioxide and nitrogen and forms after a mine explosion.

- Fire damp: consists of mostly methane, a highly flammable gas that explodes between 5% and 15% - at 25% it causes asphyxiation.

- Stink damp: so named for the rotten egg smell of the hydrogen sulphide gas, stink damp can explode and is also very toxic.

- White damp: air containing carbon monoxide which is toxic, even at low concentrations.

Safer Times in Modern Mining

Improvements in mining methods (e.g. longwall mining), hazardous gas monitoring (such as safety-lamps or more modern electronic gas monitors), gas drainage, electrical equipment, and ventilation have reduced many of the risks of rock falls, explosions, and unhealthy air quality. Gases released during the mining process can be recovered to generate electricity and improve worker safety with gas engines. Another innovation in recent years is the use of closed circuit escape respirators, respirators that contain oxygen for situations where mine ventilation is compromised. Statistical analyses performed by the US Department of Labor's Mine Safety and Health Administration (MSHA) show that between 1990 and 2004, the industry cut the rate of injuries by more than half and fatalities by two-thirds. However, according to the Bureau of Labor Statistics, even in 2006, mining remained the second most dangerous occupation in America, when measured by fatality rate. However, these numbers include all mining, with oil and gas mining contributing the majority of fatalities; coal mining resulted in only 47 fatalities that year.

Environmental Impacts

Coal mining can result in a number of adverse effects on the environment.

Surface mining of coal completely eliminates existing vegetation, destroys the genetic soil profile, displaces or destroys wildlife and habitat, degrades air quality, alters current land uses, and to some extent permanently changes the general topography of the area mined. This often results in a scarred landscape with no scenic value. Of greater concern, the movement, storage, and redistribution of soil during mining can disrupt the community of soil microorganisms and consequently nutrient cycling processes. Rehabilitation or reclamation mitigates some of these concerns and is required by US Federal Law, specifically the Surface Mining Control and Reclamation Act of 1977.

Mine dumps (tailings) could produce acid mine drainage which can seep into waterways and aquifers, with consequences on ecological and human health.

If underground mine tunnels collapse, they cause subsidence of the ground above. Subsidence can damage buildings, and disrupt the flow of streams and rivers by interfering with the natural drainage.

During actual mining operations, methane, a known greenhouse gas, may be released into the air.

Coal Mining by Country

A view of Murton colliery near Seaham, United Kingdom, 1843

Top 10 hard and brown coal producers in 2012 were (in million metric tons): China 3,621, United States 922, India 629, Australia 432, Indonesia 410, Russia 351, South Africa 261, Germany 196, Poland 144, and Kazakhstan 122.

Australia

Coal is mined in every state of Australia, but mainly in Queensland, New South Wales and Victoria. It is mostly used to generate electricity, and 75% of annual coal production is exported, mostly to eastern Asia.

In 2007, 428 million tonnes of coal was mined in Australia. In 2007, coal provided about 85% of Australia's electricity production. In fiscal year 2008/09, 487 million tonnes of coal was mined, and 261 million tonnes was exported. In fiscal year 2013/14, 430.9 million tonnes of coal was mined, and 375.1 million tonnes was exported. In 2013/14, coal provided about 69% of Australia's electricity production.

In 2013, Australia was the world's fifth-largest coal producer, after China, the United States, India, and Indonesia. However, in terms of proportion of production exported, Australia is the world's second largest coal exporter, as it exports roughly 73% of its coal production. Indonesia exports about 87% of its coal production.

Canada

Canada was ranked as the 15th coal producing country in the world in 2010, with a total production of 67.9 million tonnes. Canada's coal reserves, the 12th largest in the world, are located largely in the province of Alberta.

The first coal mines in North America were located in Joggins and Port Morien, Nova Scotia, mined by French settlers beginning in the late 1600s. The coal was used for the British garrison at Annapolis Royal, and in construction of the Fortress of Louisbourg.

Chile

China

The People's Republic of China is by far the largest producer of coal in the world, producing over 2.8 billion tons of coal in 2007, or approximately 39.8 percent of all coal produced in the world during that year. For comparison, the second largest producer, the United States, produced more than 1.1 billion tons in 2007. An estimated 5 million people work in China's coal-mining industry. As many as 20,000 miners die in accidents each year. Most Chinese mines are deep underground and do not produce the surface disruption typical of strip mines. Although there is some evidence of reclamation of mined land for use as parks, China does not require extensive reclamation and is creating significant acreages of abandoned mined land, which is unsuitable for agriculture or other human uses, and inhospitable to indigenous wildlife. Chinese underground mines often experience severe surface subsidence (6–12 meters), negatively impacting farmland because it no longer drains well. China uses some subsidence areas for aquaculture ponds but has more than they need for that purpose. Reclamation of subsided ground is a significant problem in China. Because most Chinese coal is for domestic consumption, and is burned with little or no air pollution control equipment, it contributes greatly to visible smoke and severe air pollution in industrial areas using coal for fuel. China's total energy uses 67% from coal mines.

Colombia

Opencast coal mine at Cerrejón

Some of the world's largest coal reserves are located in South America, and an opencast mine at Cerrejón in Colombia is one of the world's largest open pit mines. Output of the mine in 2004 was 24.9 million tons (compared to total global hard coal production of 4,600 million tons). Cerrejón contributed about half of Colombia's coal exports of 52 million tons that year, with Colombia ranked sixth among major coal exporting nations. The company planned to expand production to 32 million tons by 2008. The company has its own 150 km standard-gauge railroad, connecting the mine to its coal-loading terminal at Puerto Bolívar on the Caribbean coast. There are two 120-car unit trains, each carrying 12,000 tons of coal per trip. The round-trip time for each train, including loading and unloading, is about 12 hours. The coal facilities at the port are capable of loading 4,800 tons per hour on to vessels of up to 175,000 tons of dead weight. The mine, railroad and port operate 24 hours per day. Cerrejón directly employs 4,600 workers, with a further 3,800 employed by contractors. The reserves at Cerrejón are low-sulfur, low-ash, bituminous coal. The coal is mostly used for electric power generation, with some also used in steel manufacture. The surface mineable reserves for the current contract are 330 million tons. However, total proven reserves to a depth of 300 metres are 3,000 million tons.

Germany

Germany has a long history of coal mining, going back to the Middle Ages. Coal mining greatly increased during the industrial revolution and the following decades. The main mining areas were around Aachen, the Ruhr and Saar area, along with many smaller areas in other parts of Germany. These areas grew and were shaped by coal mining and coal processing, and this is still visible even after the end of the coal mining.

Coal mining reached its peak in the first half of the 20th century. After 1950, the coal producers started to struggle financially. In 1975, a subsidy was introduced (*Kohlepfennig*). In 2007, the Bundestag decided to end subsidies by 2018. As a consequence, RAG Aktiengesellschaft, the owner of the two remaining coal mines in Germany, announced it would close all mines by 2018, thus ending coal mining in Germany.

India

Coal mining in India has a long history of commercial exploitation covering nearly 220 years starting in 1774 with John Sumner and Suetonius Grant Heatly of the East India Company in the Raniganj Coalfield along the Western bank of river Damodar. However, for about a century the growth of Indian coal mining remained sluggish for want of demand but the introduction of steam locomotives in 1853 gave a fillip to it. Within a short span, production rose to an annual average of 1 million tonne (mt) and India could produce 6.12 mts. per year by 1900 and 18 mts per year by 1920. The production got a sudden boost from the First World War but went through a slump in the early thirties. The production reached a level of 29 mts. by 1942 and 30 mts. by 1946. With the advent of Independence, the country embarked upon the 5-year development plans. At the beginning of the 1st Plan, annual production went up to 33 mts. During the 1st Plan period itself, the need for increasing coal production efficiently by systematic and scientific development of the coal industry was being felt. Setting up of the National Coal Development Corporation (NCDC), a Government of India Undertaking in 1956 with the collieries owned by the railways as its nucleus was the first major step towards planned development of Indian Coal Industry. Along with the Singareni Collieries Company Ltd. (SCCL) which was already in operation since 1945 and which became a Government company under the control of Government of Andhra Pradesh in 1956, India thus had two Government coal companies in the fifties. SCCL is now a joint undertaking of Government of Telangana and Government of India sharing its equity in 51:49 ratio.

Japan

The Daikōdō (大抗道), the first adit of the Horonai mine (1879).(also known as the Otowakō (音羽坑))

The Japanese archipelago counts four main islands, the richest coal deposits have been found on the northernmost and the southernmost island: Hokkaidō and Kyūshū.

Japan has a long history of coal mining dating back into the Japanese Middle Ages. It has been told that the first coal has been discovered by a farmer couple in the region of Ōmuta, central Kyūshū in 1469. Nine years later, in 1478, local farmers discovered burning stones in the north of the Island, which meant the start of the exploitation of the Chikuhō coalfield.

The discovery of the coalfields in the north are only since the Japanese industrialization. One of the first mines in Hokkaidō was the Hokutan Horonai coal mine.

Poland

Russia

Russia was ranked as the 5th coal producing country in the world in 2010, with a total production of 316.9 million tonnes. Russia itself is the possessor of the world's second largest coal reserves. Russia also has equal rights to coal located in the Arctic archipelago of Svalbard, in accordance with the Svalbard Treaty.

Spain

Spain was ranked as the 30th coal producing country in the world in 2010. The coal miners of Spain were active in the Spanish Civil War on the Republican side. In October 1934, in Asturias, union miners and others suffered a fifteen-day siege in Oviedo and Gjion. There is a museum dedicated to coal mining in the region of Catalonia, called Cercs Mine Museum.

South Africa

South Africa is one of the ten largest coal producing and the fourth largest coal exporting country in the world.

Taiwan

Abandoned coal mine in Pingxi, New Taipei.

In Taiwan, coal is distributed mainly in the northern area. All of the commercial coal deposits occurred in three Miocene coal-bearing formations, which are the Upper, the Middle and the Lower Coal Measures. The Middle Coal Measures was the most important with its wide distribution, great number of coal beds and extensive potential reserves. Taiwan has coal reserves estimated to be 100-180 Mt. However, coal output had been small, amounting to 6,948 metric tonnes per month from 4 pits before it ceased production effectively in 2000. The abandoned coal mine in Pingxi District, New Taipei has now turned into the Taiwan Coal Mine Museum.

Ukraine

In 2012 coal production in Ukraine amounted to 85.946 million tonnes, up 4.8% from 2011. Coal consumption that same year grew by to 61.207 million tonnes, up 6.2% compared with 2011.

More than 90 percent of Ukraine's coal production comes from the Donets Basin. The country's coal industry employs about 500,000 people. Ukrainian coal mines are among the most dangerous in the world, and accidents are common. Furthermore, the country is plagued with extremely dangerous illegal mines.

United Kingdom

The United Kingdom was ranked as the 24th coal producing country in the world in 2010, with a total production of 18.2 million tonnes. Coal mining in the United Kingdom probably dates to Roman times; coal production increased significantly during the Industrial Revolution in the 19th century and peaked during World War I. As a result of its long history with coal Britain's economically recoverable coal reserves have decreased, and more than twice as much coal is now imported than produced.

United States

Miners at the Virginia-Pocahontas Coal Company Mine in 1974

The American share of world coal production remained steady at about 20 percent from 1980 to 2005, at about 1 billion short tons per year. The United States was ranked as the 2nd coal producing country in the world in 2010, and possesses the largest coal reserves in the world. In 2008 then-President George W. Bush stated that coal was the most reliable source of electricity. However, in 2011 President Barack Obama said that the US should rely more on "clean" sources of energy that emit lower or no carbon dioxide pollution. As of 2013, while domestic coal consumption for electric power was being displaced by natural gas, exports were increasing. US coal production increasingly comes from strip mines in the western United States, such as from the Powder River Basin in Wyoming and Montana.

Gold Mining

Super Pit gold mine in Western Australia

Gold-bearing quartz veins in Alaska

Gold mining is the process of mining of gold or gold ores from the ground. There are several techniques and processes by which gold may be extracted from the earth.

History

A miner underground at Pumsaint gold mine Wales; c. 1938?.

Landscape of Las Médulas, Spain, the result of hydraulic mining on a vast scale by the Ancient Romans

Late 15th and early 16th century mining techniques, *De re metallica*

It is impossible to know the exact date that humans first began to mine gold, but some of the oldest known gold artifacts were found in the Varna Necropolis in Bulgaria. The graves of the necropolis were built between 4700 and 4200 BC, indicating that gold mining could be at least 7000 years old. A group of German and Georgian archaeologists claims the Sakdrisi site in southern Georgia, dating to the 3rd or 4th millennium BC, may be the world's oldest known gold mine.

Bronze age gold objects are plentiful, especially in Ireland and Spain, and there are several well known possible sources. Romans used hydraulic mining methods, such as hushing and ground sluicing on a large scale to extract gold from extensive alluvial (loose sediment) deposits, such as those at Las Medulas. Mining was under the control of the state but the mines may have been leased to civilian contractors some time later. The gold served as the primary medium of exchange within the empire, and was an important motive in the Roman invasion of Britain by Claudius in the first century AD, although there is only one known Roman gold mine at Dolaucothi in west Wales. Gold was a prime motivation for the campaign in Dacia when the Romans invaded Transylvania in what is now modern Romania in the second century AD. The legions were led by the emperor Trajan, and their exploits are shown on Trajan's Column in Rome and the several reproductions of the column elsewhere (such as the Victoria and Albert Museum in London). Under the Eastern Roman Empire Emperor Justinian's rule, gold was mined in the Balkans, Anatolia, Armenia, Egypt, and Nubia.

In the area of the Kolar Gold Fields in Bangarpet Taluk, Kolar District of Karnataka state, India, gold was first mined prior to the 2nd and 3rd century AD by digging small pits. (Golden objects found in Harappa and Mohenjo-daro have been traced to Kolar through the analysis of impurities — the impurities include 11% silver concentration, found only in KGF ore.) The Champion reef at the Kolar gold fields was mined to a depth of 50 metres (160 ft) during the Gupta period in the fifth century AD. During the Chola period in the 9th and 10th century AD, the scale of the operation grew. The metal continued to be mined by the eleventh century kings of South India, the Vijayanagara Empire from 1336 to 1560, and later by Tipu Sultan, the king of Mysore state and the British. It is estimated that the total gold production in Karnataka to date is 1000 tons.

The mining of the Slovak deposit primarily around Kremnica was the largest of the Medieval period in Europe.

During the 19th century, numerous gold rushes in remote regions around the globe caused large migrations of miners, such as the California Gold Rush of 1849, the Victorian Gold Rush, and the Klondike Gold Rush. The discovery of gold in the Witwatersrand led to the Second Boer War and ultimately the founding of South Africa.

The Carlin Trend of Nevada, U.S., was discovered in 1961. Official estimates indicate that total world gold production since the beginning of civilization has been 4,970,000,000 troy ounces (155,000 t) and total Nevada production is three percent of that, which ranks Nevada as one of the Earth's primary gold producing regions.

Statistics

Despite the decreasing gold content of ores, the production is increasing. This can be achieved with industrial installations, and new process, like hydrometallurgy.

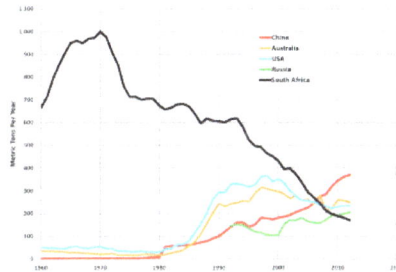

Trends in some gold-producing countries

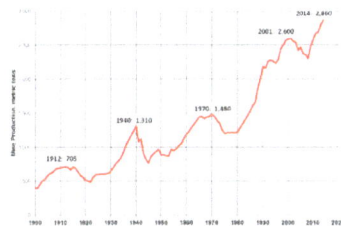

Annual world mined gold production, 1900-2014

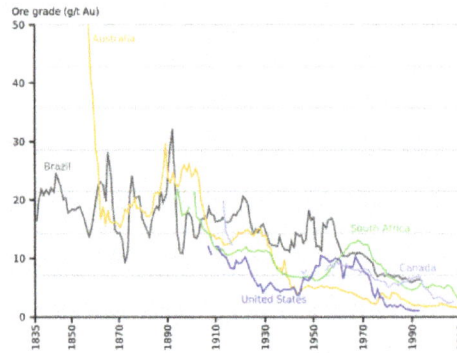

Gold ore grade evolution

Methods

Placer Mining

Placer mining is the technique by which gold that has accumulated in a placer deposit is extracted. Placer deposits are composed of relatively loose material that makes tunneling difficult, and so most means of extracting it involve the use of water or dredging.

Panning

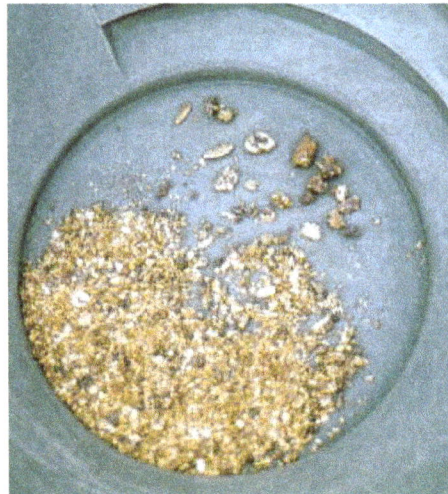

Gold in a pan—Alaska

Gold panning is mostly a manual technique of separating gold from other materials. Wide, shallow pans are filled with sand and gravel that may contain gold. The pan is submerged in water and shaken, sorting the gold from the gravel and other material. As gold is much denser than rock, it quickly settles to the bottom of the pan. The panning material is usually removed from stream beds, often at the inside turn in the stream, or from the bedrock shelf of the stream, where the density of gold allows it to concentrate, a type called placer deposits.

Gold panning is the easiest and quickest technique for searching for gold, but is not commercially viable for extracting gold from large deposits, except where labor costs are very low or gold traces are substantial. Panning is often marketed as a tourist attraction on former gold fields. Before large production methods are used, a new source must be identified and panning is useful to identify placer gold deposits to be evaluated for commercial viability.

Sluicing

Gold sluicing at Dilban Town, New Zealand, 1880s

Taking gold out of a sluice box, western North America, 1900s

Using a sluice box to extract gold from placer deposits has long been a very common practice in prospecting and small-scale mining. A sluice box is essentially a man made channel with riffles set in the bottom. The riffles are designed to create dead zones in the current to allow gold to drop out of suspension. The box is placed in the stream to channel water flow. Gold-bearing material is placed at the top of the box. The material is carried by the current through the volt where gold and other dense material settles out behind the riffles. Less dense material flows out of the box as tailings.

Larger commercial placer mining operations employ screening plants, or trommels, to remove the larger alluvial materials such as boulders and gravel, before concentrating the remainder in a sluice box or jig plant. These operations typically include diesel powered, earth moving equipment, including excavators, bulldozers, wheel loaders, and rock trucks.

Dredging

Although this method has largely been replaced by modern methods, some dredging is done by small-scale miners using suction dredges. These are small machines that float on the water and are usually operated by one or two people. A suction dredge consists of a sluice box supported by pontoons, attached to a suction hose which is controlled by a miner working beneath the water.

State dredging permits in many of the United States gold dredging areas specify a seasonal time period and area closures to avoid conflicts between dredgers and the spawning time of fish populations. Some states, such as Montana, require an extensive permitting procedure, including permits from the U.S. Corps of Engineers, the Montana Department of Environmental Quality, and the local county water quality boards.

Some large suction dredges (100 horsepower (75 kW) & 250 mm (10 in)) are used in commercial production throughout the world. Small suction dredges are much more efficient at extracting smaller gold than the old *bucket line*. This has improved the chances of finding gold. Smaller dredges with 50-to-100-millimetre (2 to 4 in) suction tubes are used to sample areas behind boulders and along potential pay streaks, until "colour" (gold) appears.

Other larger scale dredging operations take place on exposed river gravel bars at seasonal low water. These operations typically use a land based excavator to feed a gravel screening plant and sluice box floating in a temporary pond. The pond is excavated in the gravel bar and filled from the natural water table. "Pay" gravel is excavated from the front face of the pond and processed through the floating plant, with the gold trapped in the onboard sluice box and tailings stacked behind the plant, steadily filling in the back of the pond as the operation moves forward. This type of gold mining is characterized by its low cost, as each rock is moved only once. It also has low environmental impact, as no stripping of vegetation or overburden is necessary, and all process water is fully recycled. Such operations are typical on New Zealand's South Island and in the Klondike region of Canada.

Rocker Box

Thealso called a cradle, uses a riffles located in a high-walled box to trap gold in a similar manner to the sluice box. A rocker box uses less water than a sluice box and is well suited for areas where water is limited. A rocking motion provides the water movement needed for the gravity separation of gold in placer material.

Hard Rock Mining

Hard rock mining at the Associated Gold Mine, Kalgoorlie, Australia, 1951

Gold mining in Coromandel Peninsula, New Zealand in the 1890s

Hard rock gold mining extracts gold encased in rock, rather than fragments in loose sediment, and produces most of the world's gold. Sometimes open-pit mining is used, such as at the Fort Knox Mine in central Alaska. Barrick Gold Corporation has one of the largest open-pit gold mines in North America located on its Goldstrike mine property in north eastern Nevada. Other gold mines use underground mining, where the ore is extracted through tunnels or shafts. South Africa has the world's deepest hard rock gold mine up to 3,900 metres (12,800 ft) underground. At such depths, the heat is unbearable for humans, and air conditioning is required for the safety of the workers. The first such mine to receive air conditioning was Robinson Deep, at that time the deepest mine in the world for any mineral.

By-product Gold Mining

Gold is also produced by mining in which it is not the principal product. Large copper mines, such as the Bingham Canyon mine in Utah, often recover considerable amounts of gold and other metals along with copper. Some sand and gravel pits, such as those around Denver, Colorado, may recover small amounts of gold in their washing operations. The largest producing gold mine in the world, the Grasberg mine in Papua, Indonesia, is primarily a copper mine.

A modest amount of precious metal is a by-product of sodium production.

Gold Ore Processing

In placer mines, the gold is recovered by gravity separation. For hard rock mining, other methods are usually used.

Cyanide Process

Cyanide extraction of gold may be used in areas where fine gold-bearing rocks are found. Sodium cyanide solution is mixed with finely ground rock that is proven to contain gold or silver, and is then separated from the ground rock as gold cyanide or silver cyanide solution. Zinc is added to precipitate out residual zinc as well as the silver and gold metals. The zinc is removed with sulfuric acid, leaving a silver or gold sludge that is generally smelted into an ingot then shipped to a metals refinery for final processing into 99.9999% pure metals.

Advancements in the 1970s have seen activated carbon used in extracting gold from the leach solution. The gold is absorbed into the porous matrix of the carbon. Activated carbon has so much internal surface area, that fifteen grams of it has the equivalent surface area of the Melbourne Cricket Ground (18,100 square metres (195,000 sq ft)). The gold can be removed from the carbon by using a strong solution of caustic soda and cyanide, a process known as elution. Gold is then plated out onto steel wool through electrowinning. Gold specific resins can also be used in place of activated carbon, or where selective separation of gold from copper or other dissolved metals is required.

The technique using dissolution with alkaline cyanide has been highly developed over recent years. It is particularly appropriate for low grade gold and silver ore processing (e.g. less than 5 ppm gold) but its use is not restricted to such ores. There are many environmental hazards associated with this extraction method, largely due to the high acute toxicity of the cyanide compounds involved. A major example of this hazard was demonstrated in the 2000 Baia Mare cyanide spill, when a break in holding pond dam at a mine waste reprocessing facility near Baia Mare in northern Romania released approximately 100,000 cubic metres (3,500,000 cu ft) of waste water contaminated with heavy metal sludge and up to 120 long tons (122 t) of cyanide into the Tisza River. As a consequence, most countries now have strict regulations for cyanide in plant discharges, and plants today include a specific cyanide-destruction step before discharging their tailings to a storage facility.

Mercury Process

Historically, mercury was used extensively in placer gold mining in order to form mercury-gold amalgam with smaller gold particles, and thereby increase the gold recovery rates. Large-scale use of mercury stopped in the 1960s. However, mercury is still used in small scale, often clandestine, gold prospecting. It is estimated that 45,000 metric tons of mercury used in California for placer mining have not been recovered.

Business

Small Operations

Woman panning for gold in Guinea

While most of the gold is produced by major corporations, tens of thousands of people work independently in smaller, artisan operations, in some cases illegally. In Ghana, for instance, the *galamseys* are estimated to number 20,000 to 50,000. In neighboring francophone countries, such workers are called *orpailleurs*. In Brazil, such workers are called *garimpeiros*.

The high risk of such ventures was seen in the collapse of an illegal mine at Dompoase, Ashanti Region, Ghana, on 12 November 2009, when 18 workers were killed, including 13 women. Many women work at such mines as porters. It was the worst mining disaster in Ghanaian history.

In order to maximize gold extraction, mercury is often used to amalgamate with the metal. The gold is produced by boiling away the mercury from the amalgam. Mercury is effective in extracting very small gold particles, but the process is hazardous due to the toxicity of mercury vapour.

Large Companies

Barrick Gold, Goldcorp, Newmont Mining Corporation, Newcrest Mining, and AngloGold Ashanti are the world's five largest gold mining companies by market capitalisation in 2008.

Adverse Effects

Gold mining can significantly alter the natural environment. For example, gold mining activities in tropical forests are increasingly causing deforestation along rivers and in remote areas rich in biodiversity. Other gold mining impacts, particularly in aquatic systems with residual cyanide or mercury (used in the recovery of gold from ore) can be highly toxic to people and wildlife even at relatively low concentrations.

Silver Mining

An underground silver mine in Baden-Württemberg, Germany

Silver mining is the resource extraction of silver by mining.

Silver is found in native form very rarely as nuggets, but more usually combined with sulfur, arsenic, antimony, or chlorine and in various ores such as argentite (Ag_2S), chlorargyrite ("horn silver," $AgCl$), and galena (a lead ore often containing significant amounts of silver). As silver is often found in conjunction with these or alloyed with other metals such as gold, it usually must be further extracted through amalgamation or electrolysis.

Silver mining has been undertaken since early times. As silver is a precious metal often used for coins, its mining has historically often been lucrative. As with other precious metals such as gold or platinum, newly discovered deposits of silver ore have sparked silver rushes of miners seeking their fortunes. In recent centuries, large deposits were discovered and mined in the Americas, influencing the growth and development of Mexico, Andean countries such as Bolivia, Chile, and Peru, as well as Canada and the United States.

History

Early silver Athenian coin, 5th century BCE. British Museum.

Silver has been known since ancient times. Silver is mentioned in the Book of Genesis, and slag heaps found in Asia Minor and on the islands of the Aegean Sea indicate that silver was being separated from lead as early as the 4th millennium BC. The silver mines at Laurium were very rich and helped provide a currency for the economy of Ancient Athens. It involved mining the ore in underground galleries, washing the ores and smelting it to produce the metal. Elaborate washing tables still exist at the site using rain water held in cisterns and collected during the winter months.

Extraction of silver from lead ore was widespread in Roman Britain as a result of Roman mining very soon after the conquest of the first century AD.

From the mid-15th century silver began to be extracted from copper ores in massive quantities using the liquation process creating a boost to the mining and metallurgy industries of Central Europe.

Americas

Vast amounts of silver were brought into the possession of the crowns of Europe after the conquest of the Americas from the now Mexican state of Zacatecas (discovered in 1546) and Potosí (Bolivia, also discovered in 1546), which triggered a period of inflation in Europe. Silver, which was extremely valuable in China, became a global commodity, contributing to the rise of the Spanish Empire. The rise and fall of its value affected the world market.

Silver Spanish real from the silver of Potosí, Bolivia. The amount of silver mined at Potosí and other locations in the Spanish Empire led to significant inflation in Europe.

In the first half of the 19th century Chilean mining revived due to a silver rush in the Norte Chico region, leading to an increased presence of Chileans in the Atacama desert and a shift away from an agriculture based economy.

Silver mining was a driving force in the settlement of western North America, with major booms for silver and associated minerals (lead, mostly) in the galena ore silver is most commonly found in. Notable silver rushes were in Colorado; Nevada; Cobalt, Ontario; California and the Kootenay region of British Columbia; notably in the Boundary and "Silvery" Slocan. The first major silver ore deposits in the United States were discovered at the Comstock Lode in Virginia City, Nevada, in 1859.

Ore Processing

Silver is commonly extracted from ore by smelting or chemical leaching. Ore treatment by mercury amalgamation, such as in the patio process or pan amalgamation was widely used through the 1800s, but is seldom used today.

Silver is also produced during the electrolytic refining of copper and by application of the Parkes process on lead metal obtained from lead ores that contain small amounts of silver. Commercial grade fine silver is at least 99.9 percent pure silver and purities greater than 99.999 percent are available.

Production Areas

The principal sources of silver are copper, copper-nickel, gold, lead, and lead-zinc ores obtained from Canada, Mexico, Poland, Peru, Bolivia, Australia and the United States.

Mexico was the world's largest silver producer in 2014, producing 5,000 metric tons (161 million troy ounces), 18.7 percent of the 26,800 tonne (862 million troy ounce) production of the world.

Top 6 Production Stage Silver Mines		
Mine	**Country**	**2010 Production**
Cannington Silver/Lead/Zinc Mine	Australia	38.6 Moz
Fresnillo Silver Mine	Mexico	38.6 Moz
San Cristobal Polymetallic Mine	Bolivia	19.4 Moz
Antamina Copper/Zinc Mine	Peru	14.9 Moz
Rudna Copper Mine	Poland	14.9 Moz
Peñasquito Polymetallic Mine	Mexico	13.9 Moz

Top 6 Near-Term Silver Mine Development Projects Through 2015		
Project	**Country**	**Anticipated Annual Production Capacity (due within five years)**
Pascua Lama	Chile	25.0 Moz
Navidad	Argentina	15.0 Moz
Juanicipio	Mexico	14.0 Moz
Malku Khota	Bolivia	13.2 Moz
Hackett River	Canada	13.1 Moz
Corani	Peru	10.0 Moz

Placer Mining

19th-century miner pouring material from a stream bed into a rocker box, which when rocked back and forth will help to separate gold dust from the sand and gravel

Placer mining is the mining of stream bed (alluvial) deposits for minerals. This may be done by open-pit (also called open-cast mining) or by various surface excavating equipment or tunneling equipment.

Placer mining is frequently used for precious metal deposits (particularly gold) and gemstones, both of which are often found in alluvial deposits—deposits of sand and gravel in modern or ancient stream beds, or occasionally glacial deposits. The metal or gemstones, having been moved by stream flow from an original source such as a vein, are typically only a minuscule portion of the total deposit. Since gems and heavy metals like gold are considerably more dense than sand, they tend to accumulate at the base of placer deposits.

The containing material may be too loose to safely mine by tunneling, though it is possible where the ground is permanently frozen. Where water under pressure is available, it may be used to mine, move, and separate the precious material from the deposit, a method known as hydraulic mining, hydraulic sluicing or hydraulicking.

Etymology

The word placer derives from the Spanish *placer*, meaning shoal or alluvial/sand deposit, from Catalan *placer*, (shoal), from *plassa*, (place) from Medieval Latin *placea* (place) the origin word for "place" and "plaza" in English. The word in Spanish is thus ultimately derived from *placea* and refers directly to an alluvial or glacial deposit of sand or gravel.

History

Das Gerinne A.   Der Trog B.   Sein umgekehrter Boden C.   Sein sütlicher Auslaß D.
Eine eiserne Krücke E.   Die Brettchen F.   Das Wassergerinne G.
Der Sichertrog, in dem das, was sich abgesetzt hat, aufgefangen wird H.
Ein Sichertrog, in dem es verwaschen wird I.

Plate depicting placer mining from the 1556 book *De re metallica*

Placers supplied most of the gold for a large part of the ancient world. Hydraulic mining methods such as hushing were used widely by the Romans across their empire, but especially in the gold fields of northern Spain after its conquest by Augustus in 25 BC. One of the largest sites was at Las Médulas, where seven 30 mile long aqueducts were used to work the alluvial gold deposits through the first century AD. (Inclusions of platinum-group metals in a very large proportion of gold items indicate that the gold was largely derived from placer or alluvial deposits. Platinum group metals are seldom found with gold in hardrock reef or vein deposits.)

In North America, placer mining was famous in the context of several gold rushes, particularly the California Gold Rush and the Colorado Gold Rush, the Fraser Canyon Gold Rush and the Klondike Gold Rush.

Placer mining continues in many areas of the world as a source of diamonds, industrial minerals and metals, gems (in Myanmar and Sri Lanka), platinum, and of gold (in Yukon, Alaska and British Columbia).

Deposits

An area well protected from the flow of water is a great location to find gold. Gold is very dense and is often found in a stream bed. Many different gold deposits are dealt with in different ways. Placer deposits attract many prospectors because their costs are very low. There are many different places gold could be placed, such as a residual, alluvial, and a bench deposit.

Residual

Residual deposits are more common where there has been weathering on rocks and where there hasn't been water. They are deposits which have not been washed away yet or been moved. The residual usually lies at the site of the lode. This type of deposit undergoes rock weathering.

Alluvial

Alluvial or eluvial deposits sometimes have the largest gold deposit and are very common. This deposit is created when a force of nature moves or washes the gold away, but it doesn't go into a stream bed. It contains pieces of ore that have been washed away from the lode. Alluvial deposits are the most common type of placer gold. This type of deposit occurs mostly in valleys.

Bench

Bench deposits are created when gold reaches a stream bed. Gold accumulations in an old stream bed that are high are called bench deposits. They can be found on higher slopes that drain into valleys. Dry stream beds (benches) can be situated far from other water sources and can sometimes be found on mountain tops. Today, many miners focus their activities on bench deposits.

Methods

A number of methods are used to mine placer gold and gems, both in terms of extracting the minerals from the ground, and separating it from the non-gold or non-gems.

Panning

Panning for gold in Korea, c. 1900

The simplest technique to extract gold from placer ore is panning. In panning, some mined ore is placed in a large metal or plastic pan, combined with a generous amount of water, and agitated so that the gold particles, being of higher density than the other material, settle to the bottom of the pan. The lighter gangue material such as sand, mud and gravel are then washed over the side of the pan, leaving the gold behind. Once a placer deposit is located by gold panning, the miner usually shifts to equipment that can treat volumes of sand and gravel more quickly and efficiently.

Rocker

A rocker box (or "cradle") is capable of greater volume than a gold pan, and it is more portable and requires less infrastructure than a sluice box, being fed not by a sluice but by hand. The box sits on rockers, which when rocked separates out the gold, and the practice was referred to as "rocking the golden baby".

Sluice Box

Riffles in a sluice box. The small specks are gold, the larger ones are merely pebbles

A modern sluice box made of metal; in its base are the riffles used to catch gold settling to the bottom

The same principle may be employed on a larger scale by constructing a short sluice box, with barriers along the bottom called riffles to trap the heavier gold particles as water washes them and the other material along the box. This method better suits excavation with shovels or similar implements to feed ore into the device. Sluice boxes can be as short as a few feet, or more than ten feet (a common term for one that is over six feet +/- is a "Long Tom").

Dry Washing

Sluicing is only effective in areas where there is a sufficient water supply, and is impractical in arid areas. Alternative methods developed that used the blowing of air to separate out gold from sand.

Trommel

Trommel at the Potato Patch, Blue Ribbon Mine, Alaska

A trommel is composed of a slightly-inclined rotating metal tube (the 'scrubber section') with a screen at its discharge end. Lifter bars, sometimes in the form of bolted in angle iron, are attached to the interior of the scrubber section. The ore is fed into the elevated end of the trommel. Water, often under pressure, is provided to the scrubber and screen sections and the combination of water and mechanical action frees the valuable minerals from the ore. The mineral bearing ore that passes through the screen is then further concentrated in smaller devices such as sluices and jigs. The larger pieces of ore that do not pass through the screen can be carried to a waste stack by a conveyor.

Gold Dredge

The Natomas No. 6 gold dredge in operation in 1958 in Folsom, California

Large-scale sifting of placer gold from large volumes of alluvial deposits can be done by use of mechanical dredges.

Underground Mining

Miners using jets of steam to melt the permafrost in an underground gold mine

In areas where the ground is permanently frozen, such as in Siberia, Alaska, and the Yukon, placer deposits may be mined underground. As the frozen ground is otherwise too hard and firm to mine by hand, historically fires were built so as to thaw the ground before digging it. Later methods involve blasting jets of steam ("points") into the frozen deposits.

Environmental Effects

Although not required, the process water may be continuously recycled and the ore from which the sought after minerals have been extracted ("the tailings") can be reclaimed. While these recycling and reclamation processes are more common in modern placer mining operations they are still not universally done.

View of Las Médulas

In earlier times the process water was not generally recycled and the spent ore was not reclaimed. The remains of a Roman alluvial gold mine at Las Médulas are so spectacular as to justify the site being designated UNESCO World Heritage status. The methods used by the Roman miners are fully described by Pliny the Elder in his work Naturalis Historia published in about 77 AD. The author was a Procurator in the region and so probably witnessed large-scale hydraulic mining of the placer deposits there. He also added that the local lake Curacado had been heavily silted by the mining methods.

Environmental activists describe the hydraulic mining form of placer mining as environmentally destructive because of the large amounts of silt that it adds to previously clear running streams (also known as the "Dahlonega Method"). Most placer mines today use settling ponds, if only to ensure that they have sufficient water to run their sluicing operations.

In California, from 1853 to 1884, "hydraulicking" of placers removed an enormous amount of material from the gold fields, material that was carried downstream and raised the level of portions of the Central Valley by some seven feet in affected areas and settled in long bars up to 20 feet thick in parts of San Francisco Bay. The process raised an opposition calling themselves the "Anti-Debris Association". In January 1884, the North Bloomfield Mining and Gravel Company case banned the flushing of debris into streams, and the hydraulic mining mania in California's gold country came to an end.

References

- M M Postan; E Miller, The Cambridge Economic History of Europe: Trade and industry in the Middle Ages, Cambridge University Press, 28 August 1987, ISBN 0521087090

- Krah, Jaclyn; Unger, Richard L. (7 August 2013). "The Importance of Occupational Safety and Health: Making for a "Super" Workplace". National Institute for Occupational Safety and Health. Retrieved 15 January 2015.

- "September 2015 - Resources and Energy Quarterly" (PDF). Australia Office of the Chief Economist. September 2015. pp. 44, 56. Retrieved 3 October 2015.

- Matthew Brown (March 17, 2013). "Company eyes coal on Montana's Crow reservation". The San Francisco Chronicle. Associated Press. Retrieved March 18, 2013.

- "43-101 Preliminary Economic Assessment Technical Report Malku Khota" (PDF). South American Silver Corp. Retrieved July 26, 2011.

- Schmidt, Stephan. "Coal deposits of South Africa - the future of coal mining in South Africa" (PDF). Institute for Geology, Technische Universität Bergakademie Freiberg. Retrieved 14 January 2010.

- "Coal". Department of Minerals and Energy (South Africa). Archived from the original on December 2, 2009. Retrieved 14 January 2010.

Mining: Methods and Processes

The methods and processes involved in any field of study is an important component of the study. This chapter involves the reader with a better understanding on automated mining, asteroid mining, deep-sea mining, Uranium mining and bio mining. The chapter serves as a source to understand the major categories related to mining.

Automated Mining

Automated mining involves the removal of human labor from the mining process. The mining industry is in the transition towards automation. It can still require a large amount of human capital, particularly in the third world where labor costs are low so there is less incentive for increasing efficiency. Automated mining is an umbrella term that refers to two types of technology. The first type of mining automation deals with process and software automation; the second type deals with applying robotic technology to mining vehicles and equipment.

Mine Automation Software

In order to gain more control over their operations, mining companies may implement mining automation software or processes. Mine management solutions like Pitram Mining Solutions and Extreme help mine administrators organize, control, and monitor mining events throughout the site in real time. Reports generated by mine automation software allow administrators to identify productivity bottlenecks, increase accountability, and better understand return on investment.

Mining Equipment Automation

Addressing concerns about how to improve productivity and safety in the mine site, some mine companies are turning to equipment automation consisting of robotic hardware and software technologies that convert vehicles or equipment into autonomous mining units.

Mine equipment automation comes in four different forms: remote control, teleoperation, driver assist, and full automation.

Remote Control

Remote control mining equipment usually refers to mining vehicles such as excavators or bulldozers that are controlled with a handheld remote control. An operator stands in line-of-sight and uses the remote control to perform the normal vehicle functions. Because visibility and feel of the machine are heavily reduced, vehicle productivity is generally reduced as well using remote control. Remote control technology is generally used to enable mining equipment to operate in dangerous conditions such as unstable terrain, blast areas or in high risk areas of falling debris, or underground mining. Remote control technology is generally the least expensive way to automate mining equipment making it an ideal entry point for companies looking to test the viability of robotic technology in their mine.

Teleoperated Mining Equipment

Teleoperated mining equipment refers to mining vehicles that are controlled by an operator at a remote location with the use of cameras, sensors, and possibly additional positioning software. Teleoperation allows an operator to further remove themselves from the mining location and control a vehicle from a more protected environment. Joysticks or other handheld controls are still used to control the vehicle's functions, and operators have greater access to vehicle telemetry and positioning data through the teleoperation software. With the operator removed from the cab, teleoperated mining vehicles may also experience reduced productivity; however, the operator has a better vantage point than remote control from on-vehicle cameras and sensors and is further removed from potentially dangerous conditions.

Driver Assist

"Driver assist" refers to partly automated control of mining machines. Only some of the functions are automated and operator intervention is needed. Common functions include both spotting assist and collision avoidance systems.

Full Automation

"Full automation" can refer to the autonomous control of one or more mining vehicles. Robotic components manage all critical vehicle functions including ignition, steering, transmission, acceleration, braking, and implement control (i.e. blade control, dump bed control, excavator bucket and boom, etc.) without the need for operator intervention. Fully autonomous mining systems experience the most productivity gains as software controls one or more mining vehicles allowing operators to take on the role of mining facilitators, troubleshooting errors and monitoring efficiency.

Benefits

The benefits of mining equipment automation technologies are varied but may include: improved safety, better fuel efficiency, increased productivity, reduced unscheduled maintenance, improved working conditions, better vehicle utilization, and reduced driver fatigue and attrition. Automation technologies are an efficient way to mitigate the effects of widespread labor shortages for positions such as haul truck driver. In the face of falling commodity prices, many mining companies are looking for ways to dramatically reduce overhead costs while still maintaining site safety and integrity; automation may be the answer.

Drawbacks

Critics of vehicle automation often focus on the potential for robotic technology to eliminate jobs while proponents counter that while some jobs will become obsolete (normally the dirty, dangerous, or monotonous jobs), others will be created. Communities supporting underprivileged workers that rely on entry level mining positions are worried about and are calling for social responsibility as mining companies transition to automation technologies that promise to increase productivity in the face of falling commodity prices. Risk averse mining companies are also reluctant to commit large amounts of capital to an unproven technology, preferring more often to enter the automation scene at lower, more inexpensive levels such as remote control.

Companies Offering Mining Equipment Automation

There are automated mining systems available for most mining equipment.

Autonomous and automated mining technology available from Original equipment manufacturer includes:

- Haul truck - Caterpillar Inc. and Komatsu Limited both offer OEM-based autonomous hauling systems. The CAT MineStar system enables fully autonomous control of drilling, dozing, hauling, longwall, and other underground mining tasks. Komatsu's Autonomous Haul System (AHS) equips haul trucks with on-board vehicle controllers, sensors, positioning systems to enable them to coordinate in a complex mining haulage system. Each of these OEM-based systems is fully integrated with proprietary technology and are proven in some of the largest mines in the world.

- LHD (Load, Haul, Dump machine) - Atlas Copco and Sandvik both offer semi-autonomous LHD solutions that are operational in large underground mines.

- Blast Hole Drills (surface) - Atlas Copco offer autonomous blast hole drills.

- Drilling jumbo (underground) - Atlas Copco and Sandvik offer automated underground drill rigs.

- Rockbreaker - Transmin provide automated rockbreakers.

A large number of vendor-independent and third-party retrofit solutions also exist. Third party services such as RCT's ControlMaster systems, Autonomous Solutions, Inc's Nav/Mobius system, Hard-Line's Teleop Auto system, retrofit to an existing fleet, provide similar automation capabilities as OEM-based systems, but allow mining companies to remain OEM neutral.

Some companies, such as Rio Tinto Group via their Technology and Innovation group develop autonomous mining technology in-house.

Examples of Autonomous Mining Equipment

Mine of the Future

Rio Tinto Group embarked on their Mine of the Future initiative in 2008. From a control center in Perth, Rio Tinto employees operate autonomous mining equipment in Australia's remote but mineral rich Pilbara region. The autonomous mining vehicles reduce the footprint of the mining giant while improving productivity and vehicle utilization. As of June 2014, Rio Tinto's autonomous mining fleet reached the milestone of 200 million tons hauled. Rio Tinto also operate a number of autonomous blast hole drill rigs.

Bingham Canyon Mine

Located near Salt Lake City, Utah, the Bingham Canyon Mine (Kennecott Utah Copper/Rio Tinto) is one of the largest open pit mine in the world and one of the world's largest copper producers. In April 2013, the mine experienced a catastrophic landslide that halted much of the mine's operations. As part of the cleanup efforts and to improve safety, mine administrators turned to remote control excavator, dozers and teleremote blast hole drills to perform work on the highly unstable terrain areas. Robotic technology helped Kennecott to reduce the steeper, more dangerous areas of the slide to allow manned vehicles access for cleanup efforts.

Next Generation Mining

BHP have deployed a number of autonomous mining equipment as part of their Next Generation Mining program. This includes autonomous drills and autonomous trucks in the Pilbara region.

Asteroid Mining

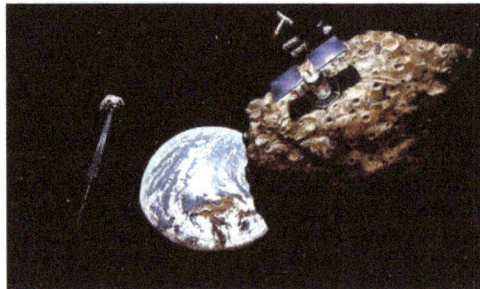

Artist's concept of asteroid mining

433 Eros is a stony asteroid in a near-Earth orbit

Asteroid mining is the exploitation of raw materials from asteroids and other minor planets, including near-Earth objects. Minerals and volatiles could be mined from an asteroid or spent comet then used in space for in-situ utilization (e.g. construction materials and rocket propellant) or taken back to Earth. These include gold, iridium, silver, osmium, palladium, platinum, rhenium, rhodium, ruthenium and tungsten for transport back to Earth; iron, cobalt, manganese, molybdenum, nickel, aluminium, and titanium for construction; water and oxygen to sustain astronauts; as well as hydrogen, ammonia, and oxygen for use as rocket propellant.

Due to the astronomically high costs of current space transportation, extraction techniques still being developed and lingering uncertainties about target selection, terrestrial mining is currently the only means of raw mineral acquisition today.

Purpose

Based on known terrestrial reserves, and growing consumption in both developed and developing countries, key elements needed for modern industry and food production could be exhausted on Earth within 50–60 years. These include phosphorus, antimony, zinc, tin, lead, indium, silver, gold and copper. In response, it has been suggested that platinum, cobalt and other valuable elements from asteroids may be mined and sent to Earth for profit, used to build solar-power satellites and space habitats, and water processed from ice to refuel orbiting propellant depots.

Although asteroids and Earth accreted from the same starting materials, Earth's relatively stronger gravity pulled all heavy siderophilic (iron-loving) elements into its core during its molten youth more than four billion years ago. This left the crust depleted of such valuable elements until a rain of asteroid impacts re-infused the depleted crust with metals like gold, cobalt, iron, manganese, molybdenum, nickel, osmium, palladium, platinum, rhenium, rhodium, ruthenium and tungsten (some flow from core to surface does occur, e.g. at the Bushveld Igneous Complex, a famously rich source of platinum-group metals). Today, these metals are mined from Earth's crust, and they are essential for economic and technological progress. Hence, the geologic history of Earth may very well set the stage for a future of asteroid mining.

In 2006, the Keck Observatory announced that the binary Jupiter trojan 617 Patroclus, and possibly large numbers of other Jupiter trojans, are likely extinct comets and consist largely of water ice. Similarly, Jupiter-family comets, and possibly near-Earth asteroids that are extinct comets, might also provide water. The process of in-situ resource utilization—using materials native to space for propellant, thermal management, tankage, radiation shielding, and other high-mass components of space infrastructure—could lead to radical reductions in its cost. Although whether these cost reductions could be achieved, and if achieved would offset the enormous infrastructure investment required, is unknown.

Ice would satisfy one of two necessary conditions to enable "human expansion into the Solar System" (the ultimate goal for human space flight proposed by the 2009 "Augustine Commission" Review of United States Human Space Flight Plans Committee): physical sustainability and economic sustainability.

From the astrobiological perspective, asteroid prospecting could provide scientific data for the search for extraterrestrial intelligence (SETI). Some astrophysicists have suggested that if advanced extraterrestrial civilizations employed asteroid mining long ago, the hallmarks of these activities might be detectable. Why extraterrestrials would have resorted to asteroid mining in near proximity to earth, with its readily available resources, has not been explained.

Asteroid Selection

Comparison of delta-v requirements for standard Hohmann transfers	
Mission	**Δv**

Earth surface to LEO	8.0 km/s
LEO to near-Earth asteroid	5.5 km/s[note 1]
LEO to lunar surface	6.3 km/s
LEO to moons of Mars	8.0 km/s

An important factor to consider in target selection is orbital economics, in particular the change in velocity (Δv) and travel time to and from the target. More of the extracted native material must be expended as propellant in higher Δv trajectories, thus less returned as payload. Direct Hohmann trajectories are faster than Hohmann trajectories assisted by planetary and/or lunar flybys, which in turn are faster than those of the Interplanetary Transport Network, but the reduction in transfer time comes at the cost of increased Δv requirements.

Near-Earth asteroids are considered likely candidates for early mining activity. Their low Δv makes them suitable for use in extracting construction materials for near-Earth space-based facilities, greatly reducing the economic cost of transporting supplies into Earth orbit.

The table above shows a comparison of Δv requirements for various missions. In terms of propulsion energy requirements, a mission to a near-Earth asteroid compares favorably to alternative mining missions.

An example of a potential target for an early asteroid mining expedition is 4660 Nereus, expected to be mainly enstatite. This body has a very low Δv compared to lifting materials from the surface of the Moon. However it would require a much longer round-trip to return the material.

Multiple types of asteroids have been identified but the three main types would include the C-type, S-type, and M-type asteroids:

1. C-type asteroids have a high abundance of water which is not currently of use for mining but could be used in an exploration effort beyond the asteroid. Mission costs could be reduced by using the available water from the asteroid. C-type asteroids also have a lot of organic carbon, phosphorus, and other key ingredients for fertilizer which could be used to grow food.

2. S-type asteroids carry little water but look more attractive because they contain numerous metals including: nickel, cobalt and more valuable metals such as gold, platinum and rhodium. A small 10-meter S-type asteroid contains about 650,000 kg (1,433,000 lb) of metal with 50 kg (110 lb) in the form of rare metals like platinum and gold.

3. M-type asteroids are rare but contain up to 10 times more metal than S-types.

A class of *easily recoverable objects* (EROs) was identified by a group of researchers in 2013. Twelve asteroids made up the initially identified group, all of which could be potentially mined with present-day rocket technology. Of 9,000 asteroids searched in the NEO database, these twelve could all be brought into an Earth-accessible orbit by changing their velocity by less than 500 meters per second (1,800 km/h; 1,100 mph). The dozen asteroids range in size from 2 to 20 meters (10 to 70 ft). Many authors have pointed out, however, the human error or technological failure might alter asteroid orbits to create disastrous asteroid strikes.

Asteroid Cataloging

The B612 Foundation is a private nonprofit foundation with headquarters in the United States, dedicated to protecting Earth from asteroid strikes. As a non-governmental organization it has conducted two lines of related research to help detect asteroids that could one day strike Earth, and find the technological means to divert their path to avoid such collisions.

The foundation's current goal is to design and build a privately financed asteroid-finding space telescope, Sentinel, to be launched in 2017–2018. The Sentinel's infrared telescope, once parked in an orbit similar to that of Venus, will help identify threatening asteroids by cataloging 90% of those with diameters larger than 140 metres (460 ft), as well as surveying smaller Solar System objects.

Data gathered by Sentinel will be provided through an existing scientific data-sharing network that includes NASA and academic institutions such as the Minor Planet Center in Cambridge, Massachusetts. Given the satellite's telescopic accuracy, Sentinel's data may prove valuable for other possible future missions, such as asteroid mining.

Mining Considerations

There are three options for mining:

1. Bring raw asteroidal material to Earth for use.

2. Process it on-site to bring back only processed materials, and perhaps produce propellant for the return trip.

3. Transport the asteroid to a safe orbit around the Moon, Earth or to the ISS. This can hypothetically allow for most materials to be used and not wasted. Along these lines, NASA has proposed a potential future space mission known as the Asteroid Redirect Mission, although the primary focus of this mission is on retrieval. The House of Representatives recently deleted a line item for the ARP budget from NASA's FY 2017 budget request.

Processing *in situ* for the purpose of extracting high-value minerals will reduce the energy requirements for transporting the materials, although the processing facilities must first be transported to the mining site.

Mining operations require special equipment to handle the extraction and processing of ore in outer space. The machinery will need to be anchored to the body, but once in place, the ore can be moved about more readily due to the lack of gravity. However, no techniques for refining oar in zero gravity currently exist. Docking with an asteroid might be performed using a harpoon-like process, where a projectile would penetrate the surface to serve as an anchor; then an attached cable would be used to winch the vehicle to the surface, if the asteroid is both penetrable and rigid enough for a harpoon to be effective.

Due to the distance from Earth to an asteroid selected for mining, the round-trip time for communications will be several minutes or more, except during occasional close approaches to Earth by near-Earth asteroids. Thus any mining equipment will either need to be highly automated, or a human presence will be needed nearby. Humans would also be useful for troubleshooting problems and for maintaining the equipment. On the other hand, multi-minute communications delays have not prevented the success of robotic exploration of Mars, and automated systems would be much less expensive to build and deploy.

Technology being developed by Planetary Resources to locate and harvest these asteroids has resulted in the plans for three different types of satellites:

1. Arkyd Series 100 (The Leo Space telescope) is a less expensive instrument that will be used to find, analyze, and see what resources are available on nearby asteroids.

2. Arkyd Series 200 (The Interceptor) Satellite that would actually land on the asteroid to get a closer analysis of the available resources.

3. Arkyd Series 300 (Rendezvous Prospector) Satellite developed for research and finding resources deeper in space.

Technology being developed by Deep Space Industries to examine, sample, and harvest asteroids is divided into three families of spacecrafts:

1. FireFlies are triplets of nearly identical spacecraft in CubeSat form launched to different asteroids to rendezvous and examine them.

2. DragonFlies also are launched in waves of three nearly identical spacecraft to gather small samples (5–10 kg) and return them to Earth for analysis.

3. Harvestors voyage out to asteroids to gather hundreds of tons of material for return to high Earth orbit for processing.

Asteroid mining could potentially revolutionize space exploration. The C-type asteroids's high abundance of water could be used to produce fuel by splitting water into hydrogen and oxygen. This would make space travel a more feasible option by lowering cost of fuel, although cost of fuel is a relatively insignificant factor in the overall cost of a manned space mission.

Extraction Techniques

Surface Mining

On some types of asteroids, material may be scraped off the surface using a scoop or auger, or for larger pieces, an "active grab." There is strong evidence that many asteroids consist of rubble piles, making this approach possible.

Shaft Mining

A mine can be dug into the asteroid, and the material extracted through the shaft. This requires precise knowledge to engineer accuracy of astro-location under the surface regolith and a transportation system to carry the desired ore to the processing facility.

Magnetic Rakes

Asteroids with a high metal content may be covered in loose grains that can be gathered by means of a magnet.

Heating

For volatile materials in extinct comets, heat can be used to melt and vaporize the matrix.

Extraction Using the Mond Process

The nickel and iron of an iron rich asteroid could be extracted by the Mond process. This involves passing carbon monoxide over the asteroid at a temperature between 50 and 60°C, then nickel and iron can be removed from the gas again at higher temperatures, perhaps in an attached printer, and platinum, gold etc. left as a residue.

Self-replicating Machines

A 1980 NASA study entitled *Advanced Automation for Space Missions* proposed a complex automated factory on the Moon that would work over several years to build a copy of itself. Exponential growth of factories over many years could refine large amounts of lunar (or asteroidal) regolith. Since 1980 there has been major progress in miniaturization, nanotechnology, materials science, and additive manufacturing, so the self-replicating "factory" might be as small as a 3-D printer.

Proposed Mining Projects

On April 24, 2012 a plan was announced by billionaire entrepreneurs to mine asteroids for their resources. The company is called Planetary Resources and its founders include aerospace entrepreneurs Eric Anderson and Peter Diamandis. Advisers include film director and explorer James Cameron and investors include Google's chief executive Larry Page and its executive chairman Eric Schmidt. They also plan to create a fuel depot in space by 2020 by using water from asteroids, splitting it to liquid oxygen and liquid hydrogen for rocket fuel. From there, it could be shipped to Earth orbit for refueling commercial satellites or spacecraft. The plan has been met with skepticism by some scientists, who do not see it as cost-effective, even though platinum and gold are worth nearly £35 per gram (approximately $1,800 per troy ounce). Platinum and gold are raw materials traded on terrestrial markets, and it is impossible to predict what prices either will command at the point in the future when resources from asteroids become available. For example, platinum, which was trading at $1800/ounce 9 years ago trades in a range between $900/1000/ounvr currently, and since the primary use of platinum is as the catalyzt in catalytic converters from internal combustion engine exhaust, the long term demand for platinum may well decrease. The upcoming NASA mission OSIRIS-REx, which is planned to return just a minimum amount (60 g; two ounces) of material but could get up to 2 kg from an asteroid to Earth, will cost about US$1 billion.

Planetary Resources says that, in order to be successful, it will need to develop technologies that bring the cost of space flight down. Planetary Resources also expects that the construction of "space infrastructure" will help to reduce long-term running costs. For example, fuel costs can be reduced by extracting water from asteroids and split it to hydrogen using solar energy. In theory, hydrogen fuel mined from asteroids costs significantly less than fuel from Earth due to high costs of escaping Earth's gravity. If successful, investment in "space infrastructure" and economies of scale could reduce operational costs to levels significantly below NASA's upcoming (OSIRIS-REx) mission.This investment would have to be amortized through the sale of commodities, delaying and/eliminating any return to investors. There are some indications that Planetary Resources expects government to fund infrastructure development, viz. its resent request for $700,000. from NASA to fund the first of the telescopes described above. The British Company, Asteroid Mining Corporation, has already announced its plans to seek government funding.

Another similar venture, called Deep Space Industries, was started by David Gump, who had founded other space companies. The company hopes to begin prospecting for asteroids suitable for mining by 2015 and by 2016 return asteroid samples to Earth. By 2023 Deep Space Industries plans to begin mining asteroids.

At ISDC-San Diego 2013, Kepler Energy and Space Engineering (KESE,llc) also announced it was going to mine asteroids, using a simpler, more straightforward approach: KESE plans to use almost exclusively existing guidance, navigation and anchoring technologies from mostly successful missions like the Rosetta/Philae, Dawn, and Hyabusa's Muses-C and current NASA Technology Transfer tooling to build and send a 4-module Automated Mining System (AMS) to a small asteroid with a simple digging tool to collect ~40 tons of asteroid regolith and bring each of the four return modules back to low Earth orbit (LEO) by the end of the decade. Small asteroids are expected to be loose piles of rubble, therefore providing for easy extraction.

In September 2012, the NASA Institute for Advanced Concepts (NIAC) announced the Robotic Asteroid Prospector project, which will examine and evaluate the feasibility of asteroid mining in terms of means, methods, and systems.

In February 2016, the British-based Asteroid Mining Corporation was established by Mitch Hunter-Scullion with the intentions of lobbying the British Government for a regulatory framework and start up investment in Asteroid Mining. Mission plans and potential system usages are being designed currently with future plans aiming to use a prospecting satellite launched aboard a reusable Falcon 9 from SpaceX or by Skylon when it becomes operational to rendezvous with a near-Earth object and collect several kilograms of Platinum group materials which will then be returned to low Earth orbit and recovered by at a later date to be sold on at a premium. The Asteroid Mining Corporation aims to raise funds through crowdfunding, in a radically different and novel approach in industrial financing to allow a wide cross section of society to benefit from the riches of space, to this end an Indiegogo appeal is being launched on July 12, 2016.

Being the largest body in the asteroid belt, Ceres could become the main base and transport hub for future asteroid mining infrastructure, allowing mineral resources to be transported to Mars, the Moon, and Earth. Because of its small escape velocity combined with large amounts of water ice, it also could serve as a source of water, fuel, and oxygen for ships going through and beyond the asteroid belt. Transportation from Mars or the Moon to Ceres would be even more energy-efficient than transportation from Earth to the Moon.

Potential Targets

According to the Asterank database, following asteroids are best targets for mining if maximum cost-effectiveness is to be achieved:

Asteroid	Est. Value ($)	Est. Profit ($)	Δv (km/s)	Composition
Ryugu	95 billion	35 billion	4.663	Nickel, iron, cobalt, water, nitrogen, hydrogen, ammonia
1989 ML	14 billion	4 billion	4.888	Nickel, iron, cobalt
Nereus	5 billion	1 billion	4.986	Nickel, iron, cobalt
Didymos	84 billion	22 billion	5.162	Nickel, iron, cobalt
2011 UW158	8 billion	2 billion	5.187	Platinum, nickel, iron, cobalt
Anteros	5570 billion	1250 billion	5.439	magnesium silicate, aluminum, iron silicate
2001 CC21	147 billion	30 billion	5.636	magnesium silicate, aluminum, iron silicate
1992 TC	84 billion	17 billion	5.647	Nickel, iron, cobalt
2001 SG10	4 billion	0.6 billion	5.880	Nickel, iron, cobalt
2002 DO3	0.3 billion	0.06 billion	5.894	Nickel, iron, cobalt

Economics and Safety

Currently, the quality of the ore and the consequent cost and mass of equipment required to extract it are unknown and can only be speculated. Some economic analyses indicate that the cost of returning asteroidal materials to Earth far outweighs their market value, and that asteroid mining will not attract private investment at current commodity prices and space transportation costs. Other studies suggest large profit by using solar power. Potential markets for materials can be identified and profit generated if extraction cost is brought down. For example, the delivery of multiple tonnes of water to low Earth orbit for rocket fuel preparation for space tourism could generate a significant profit if space tourism itself proves profitable, which has not been proven.

In 1997 it was speculated that a relatively small metallic asteroid with a diameter of 1.6 km (1 mi) contains more than US$20 trillion worth of industrial and precious metals. A comparatively small M-type asteroid with a mean diameter of 1 km (0.62 mi) could contain more than two billion metric tons of iron–nickel ore, or two to three times the world production of 2004. The asteroid 16 Psyche is believed to contain $1.7×10^{19}$ kg of nickel–iron, which could supply the world production requirement for several million years. A small portion of the extracted material would also be precious metals.

Not all mined materials from asteroids would be cost-effective, especially for the potential return of economic amounts of material to Earth. For potential return to Earth, platinum is considered very rare in terrestrial geologic formations and therefore is potentially worth bringing some quantity for terrestrial use. However, platinum from asteroids would have to be processed in orbit, since it requires 20 tons of high grade platinum oar - the equivalent of a Shuttle load - to produce an ounce of refind platinum worth +- $1000. The cost of refining in orbit is unknown, but undoubtedly man multiples of mining/refining costs within the atmosphere. Nickel, on the other hand, is quite abundant and being mined in many terrestrial locations, so the high cost of asteroid mining may not make it economically viable.

Although Planetary Resources says platinum from a 30-meter-long (98 ft) asteroid is worth US$25–50 billion, an economist remarked any outside source of precious metals could lower prices sufficiently to possibly doom the venture by rapidly increasing the available supply of such metals.

Development of an infrastructure for altering asteroid orbits could offer a large return on investment. However, astrophysicists Carl Sagan and Steven J. Ostro raised the concern altering the trajectories of asteroids near Earth may pose a collision hazard. They concluded orbit engineering has both opportunities and dangers: If controls instituted on orbit-manipulation technology were too tight, future spacefaring could be hampered, but if they were too loose, human civilization would be at risk.

Scarcity

Scarcity is a fundamental economic problem of humans having seemingly unlimited wants in a world of limited resources. Since Earth's resources are not infinite, the relative abundance of asteroidal ore gives asteroid mining the potential to provide nearly unlimited resources, which could practically eliminate scarcity for those materials.

The idea of exhausting resources is not new. In 1798, Thomas Malthus wrote, because resources are ultimately limited, the exponential growth in a population would result in falls in income per capita until poverty and starvation would result as a constricting factor on population. It should be noted that 1798 is 318 years ago, and no sign has emeged of the Malthus affect regarding raw materials.

- Proven reserves are deposits of mineral resources that are already discovered and known to be economically extractable under present or similar demand, price and other economic and technological conditions.

- Conditional reserves are discovered deposits that are not yet economically viable.

- Indicated reserves are less intensively measured deposits whose data is derived from surveys and geological projections. Hypothetical reserves and speculative resources make up this group of reserves. Inferred reserves are deposits that have been located but not yet exploited.

Continued development in asteroid mining techniques and technology will help to increase mineral discoveries. As the cost of extracting mineral resources, especially platinum group metals, on Earth rises, the cost of extracting the same resources from celestial bodies declines due to technological innovations around space exploration. However, it should be noted that the "substitution effect", i.e. the use of other materials for the functions now performed by platinum, would increase in strength as the cost of platinum increased. New supplies would also come to market in the form of jewelry and recycled electronic equipment from itinerant "we buy platinum" businesses like the "we buy gold" businesses that exist now.

Financial Feasibility

Space ventures are high-risk, with long lead times and heavy capital investment, and that is no different for asteroid-mining projects. These types of ventures could be funded through private investment or through government investment. For a commercial venture it can be profitable as long as the revenue earned is greater than total costs (costs for extraction and costs for marketing). The costs involving an asteroid-mining venture have been estimated to be around $100 billion US.

There are six categories of cost considered for an asteroid mining venture:

1. Research and development costs

2. Exploration and prospecting costs

3. Construction and infrastructure development costs

4. Operational and engineering costs

5. Environmental costs

6. Time cost

Determining financial feasibility is best represented through net present value. One requirement needed for financial feasibility is a high return on investments estimating around 30%. Example calculation assumes for simplicity that the only valuable material on asteroids is platinum. On September 5, 2008 platinum was valued at US\$1,340 per ounce, or US\$43,000 per kilogram. On August 16, 2016 is \$1157. or \$37,000 per kilogram. At the \$1,340. price, for a 10% return on investment, 173,400 kg (5,575,000 ozt) of platinum would have to be extracted for every 1,155,000 tons of asteroid ore. For a 50% return on investment 1,703,000 kg (54,750,000 ozt) of platinum would have to be extracted for every 11,350,000 tons of asteroid ore. This analysis assumes that doubling the supply of platinum to the market (5.13 million ounces in 2014) would have no affect on the price of platinum. A more realistic assumption is that increasing the supply by this amount would reduce the price 30-50%.

Regulation

Space law involves a specific set of international treaties, along with national commercialization laws. The system and framework for international and domestic laws were established through the United Nations Office for Outer Space Affairs. The rules, terms and agreements that considered by space law authorities to be part of the active body of international space law are the five international space treaties and five UN declarations. Approximately 100 nations and institutions were involved in negotiations. The space treaties cover many major issues such as arms control, non-appropriation of space, freedom of exploration, liability for damages, safety and rescue of astronauts and spacecraft, prevention of harmful interference with space activities and the environment, notification and registration of space activities, and the settlement of disputes. In exchange for assurances from the space power, the nonspacefaring nations acquiesced to U.S. and Soviet proposals to treat outer space as a commons (res communis) territory which belonged to no one state.

Asteroid mining in particular is regulated, among others, by the Outer Space Treaty and the Moon Agreement.

Varying degrees of criticism exist regarding international space law. Some critics accept the Outer Space Treaty, but reject the Moon Agreement. Therefore, it is important to note that even the Moon Agreement with its common heritage of mankind clause, allows space mining, extraction, private property rights and exclusive ownership rights over natural outer space resources, if removed from their natural place. The Outer Space Treaty and the Moon Agreement allow private property rights for outer space natural resources once removed from the surface, subsurface or subsoil of the moon and other celestial bodies in outer space. Thus, international space law is capable of managing newly emerging space mining activities, private space transportation, commercial spaceports and commercial space stations/habitats/settlements. Space mining involving the extraction and removal of natural resources from their natural location is without question allowable under the Outer Space Treaty and the Moon Agreement. Once removed, those natural resources can be reduced to possession, sold, traded and explored or used for scientific purposes. International space law allows space mining, specifically the extraction of natural resources. It is generally understood within the space law authorities that extracting space resources is allowable, even by private companies for profit. However, international space law prohibits property rights over territories and outer space land.

The Outer Space Treaty

After ten years of negotiations between nearly 100 nations, the Outer Space Treaty opened for signature on January 27, 1966. It entered into force as the constitution for outer space on October 10, 1967. The Outer Space Treaty was well received; it was ratified by ninety-six nations and signed by an additional twenty-seven states. The outcome has been that the basic foundation of international space law consists of five (arguably four) international space treaties, along with various written resolutions and declarations. The main international treaty is the Outer Space Treaty of 1967; it is generally viewed as the "Constitution" for outer space. By ratifying the Outer Space Treaty of 1967, ninety-eight nations agreed that outer space would belong to the "province of mankind", that all nations would have the freedom to "use" and "explore" outer space, and that both these provisions must be done in a way to "benefit all mankind." The province of mankind principle and the other key terms have not yet been specifically defined (Jasentuliyana, 1992). Critics have complained that the Outer Space Treaty is vague. Yet, international space law has worked well and has served space commercial industries and interests for many decades. The taking away and extraction of Moon rocks, for example, has been treated as being legally permissible.

The framers of Outer Space Treaty initially focused on solidifying broad terms first, with the intent to create more specific legal provisions later (Griffin, 1981: 733-734). This is why the members of the COPUOS later expanded the Outer Space Treaty norms by articulating more specific understandings which are found in the "three supplemental agreements" – The Rescue and Return Agreement of 1968, the Liability Convention of 1973, and the Registration Convention of 1976 (734).

Hobe (2006) explains that the Outer Space Treaty "explicitly and implicitly prohibits only the acquisition of territorial property rights" – public or private, but extracting space resources is allowable.

The Moon Agreement

The Moon Agreement (1979-1984) is often treated as though it is not a part of the body of international space law, and there has been extensive debate on whether or not the Moon Agreement is a valid part of international law. It entered into force in 1984, because of a five state ratification consensus procedure, agreed upon by the members of the United Nations Committee on Peaceful Uses of Outer Space (COPUOS). Still today very few nations have signed and/or ratified the Moon Agreement. In recent years this figure has crept up to a few more than a dozen nations who have signed and ratified the treaty. The other three outer space treaties experienced a high level of international cooperation in terms of signage and ratification, but the Moon Treaty went further than them, by defining the Common Heritage concept in more detail and by imposing specific obligations on the parties engaged in the exploration and/or exploitation of outer space. The Moon Treaty explicitly designates the Moon and its natural resources as part of the Common Heritage of Mankind.

After The Rescue and Return Agreement of 1968, the Liability Convention of 1973, and the Registration Convention of 1976 (734) were enacted, key actors involved in space law negotiations, set out to establish and confirm a few more legal norms which were to be embodied in the Moon Agreement, since important issues such as the environment, public health and sharing to benefit all mankind were left open. Many of the terms written into the Moon Treaty were sticking points during early negotiations.

The Moon Agreement allows space mining, specifically the extraction of natural resources. The treaty specifically provides in Article 11, paragraph 3 that:

Neither the surface nor the subsurface of the Moon, nor any part thereof or natural resources in place [emphasis added], shall become property of any State, international intergovernmental or non-governmental organization, national organization or non-governmental entity or of any natural person. The placement of personnel, space vehicles, equipment, facilities, stations and installations on or below the surface of the Moon, including structures connected with its surface or subsurface, shall not create a right of ownership over the surface or the subsurface of the Moon or any areas thereof.

This provision was negotiated into the Moon Agreement by the United States in order to make sure that natural resources extracted from the Moon were legally permissible to take. Taking natural resources out of their location, from the surface or subsurface, has been interpreted by space law authorities as meaning that those resources are no longer tied to the "in place" restrictions against ownership.

Christol (1980) in *The Moon Treaty: Fact and Fiction* explains this legal distinction. He states that the Moon Treaty " ... does allow for the removal from the Moon and other celestial bodies of their natural resources".

Legal Regimes of Some Countries

Some nations are beginning to promulgate legal regimes for extraterrestrial resource extraction. For example, the United States "SPACE Act of 2015"—facilitating private development of space resources consistent with US international treaty obligations—passed the US House of Representatives in July 2015. In November 2015 it passed the United States Senate. On 25 November US-President Barack Obama signed the *H.R.2262 - U.S. Commercial Space Launch Competitiveness Act* into law. The law recognizes the right of U.S. citizens to own space resources they obtain and encourages the commercial exploration and utilization of resources from asteroids. According to the article § 51303 of the law:

A United States citizen engaged in commercial recovery of an asteroid resource or a space resource under this chapter shall be entitled to any asteroid resource or space resource obtained, including to possess, own, transport, use, and sell the asteroid resource or space resource obtained in accordance with applicable law, including the international obligations of the United States.

In February 2016, the Government of Luxembourg announced that it would attempt to "jumpstart an industrial sector to mine asteroid resources in space" by, among other things, creating a "legal framework" and regulatory incentives for companies involved in the industry. By June 2016, announced that it would "invest more than US$200 million in research, technology demonstration, and in the direct purchase of equity in companies relocating to Luxembourg."

Missions

Ongoing and Planned

- OSIRIS-REx - planned NASA asteroid sample return mission (launch in September 2016)

- Hayabusa 2 - ongoing JAXA asteroid sample return mission (arriving at the target in 2018)

- Asteroid Redirect Mission - potential future space mission proposed by NASA (if funded, the mission would be launched in December 2020)

- Fobos-Grunt 2 - planned Roskosmos sample return mission to Phobos (launch in 2024)

Completed

First successful missions by country:

Nation	Flyby	Orbit	Landing	Sample return
USA	ICE (1985)	NEAR (1997)	NEAR (2001)	Stardust (2006)
Japan	Suisei (1986)	Hayabusa (2005)	Hayabusa (2005)	Hayabusa (2010)
EU	ICE (1985)	Rosetta (2014)	Rosetta (2014)	
USSR	Vega 1 (1986)			
China	Chang'e 2 (2012)			

In Fiction

The first mention of asteroid mining in science fiction is apparently Garrett P. Serviss' story *Edison's Conquest of Mars*, New York Evening Journal, 1898.

The 1979 film Alien, directed by Ridley Scott, is about the crew of the *Nostromo*, a commercially operated spaceship on a return trip to Earth hauling a refinery and 20 million tons of mineral ore mined from an asteroid. C. J. Cherryh's novel, *Heavy Time* focuses on the plight of asteroid miners in the Alliance-Union universe, while *Moon* is a 2009 British science fiction drama film depicting a lunar facility that mines the alternative fuel helium-3 needed to provide energy on Earth. It was notable for its realism and drama, winning several awards internationally.

In several science fiction video games, asteroid mining is a possibility. For example, in the space-MMO, EVE Online, asteroid mining is a very popular career, owing to its simplicity.

In Star Citizen, the mining occupation supports a variety of dedicated specialists, each of which has a critical role to play in the effort.

Borehole Mining

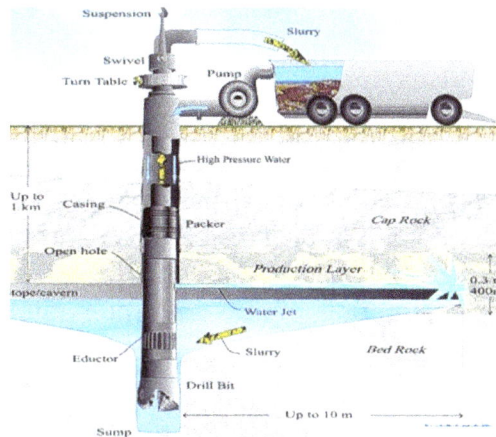

Borehole mining tool and technology principle schematic

Borehole Mining (BHM) is a remote operated method of extracting (mining) mineral resources through boreholes by means of high pressure water jets. This process can be carried-out from land surface, open pit floor, underground mine or floating platform or vessel through pre-drilled boreholes.

The Process

1. A borehole is drilled from land surface to a desired depth, where then actual borehole mining will take place.

2. A casing column is lowered down the hole. Since BHM takes place in an "open hole", the casing shoe is located just above the upper border of the production interval (ore body) leaving the rest open.

3. The BHM tool is lowered into the hole.

Description of a BHM Tool

The tool consists of at least two concentric pipes which are forming two hydraulic channels - one for pumping down a high-pressure working agent (water) and second for delivering pregnant slurry back to the surface. A BHM tool usually has (down-up): an eductor (waterjet pump) section, a hydromonitor section, an extension section and a hub, connecting it all to a drill pipe string. This string extends the tool up to the surface. Above the surface, the tool has a swivel allowing its suspension and rotation in a hole, and also connections to the working agent supply (pump station) and a slurry collector. A drill rig is normally required to operate a BHM tool.

The tool is lowered into a well until the hydromonitor reaches the required depth where the actual borehole mining is started. Then the high-pressure water is pumped down and receives back productive slurry. In the collecting pond or tank, slurry is separated and clarified water is pumped down for re-circulating.

While extracting of material, different shape underground caverns could be created. Their shapes depend on the BHM tool manipulation while mining, which obviously consist of the tool rotation, sliding it up and down and combination of these two movements. Borehole mining is applied from vertical, horizontal and deviated wells.

Advantages of BHM

The main advantages of BHM include its low capital cost, mobility, selectivity, ability to work in hazardous and dangerous conditions and low environmental impact. The method has been used in mining of such natural resources and industrial materials as: uranium, iron ore, quartz sand, gravel, coal, poly-metallic ores, phosphate, gold, diamonds, rare earths, amber and several more. Borehole mining is also used in exploration, oil, gas and water stimulation, underground storage construction and drainage.

Deep Sea Mining

Deep sea mining is a relatively new mineral retrieval process that takes place on the ocean floor. Ocean mining sites are usually around large areas of polymetallic nodules or active and extinct hydrothermal vents at about 1,400 – 3,700 m below the ocean's surface. The vents create sulfide deposits, which contain valuable metals such as silver, gold, copper, manganese, cobalt, and zinc. The deposits are mined using either hydraulic pumps or bucket systems that take ore to the surface to be processed. As with all mining operations, deep sea mining raises questions about potential environmental impact on surrounding areas. Environmental advocacy groups such as Greenpeace and the Deep sea Mining Campaign have argued that seabed mining should not be permitted in most of the world's oceans because of the potential for damage to deepsea ecosystems and pollution by heavy metal laden plumes.

Brief History

In the mid 1960s the prospect of deep-sea mining was brought up by the publication of J. L. Mero's *Mineral Resources of the Sea*. The book claimed that nearly limitless supplies of cobalt, nickel and other metals could be found throughout the planet's oceans. Mero stated that these metals occurred in deposits of manganese nodules, which appear as lumps of compressed sediment on the sea floor at depths of about 5,000 m. Some nations including France, Germany and the United States sent out research vessels in search of nodule deposits. Initial estimates of deep sea mining viability turned out to be much exaggerated. This overestimate, coupled with depressed metal prices, led to the near abandonment of nodule mining by 1982. From the 1960s to 1984 an estimated US $650 million had been spent on the venture, with little to no return.

Over the past decade a new phase of deep-sea mining has begun. Rising demand for precious metals in Japan, China, Korea and India has pushed these countries in search of new sources. Interest has recently shifted toward hydrothermal vents as the source of metals instead of scattered nodules. The trend of transition towards an electricity-based information and transportation infrastructure currently seen in western societies further pushes demands for precious metals. The current revived interest in phosphorus nodule mining at the seafloor stems from phosphor-based artificial fertilizers being of significant importance for world food production. Growing world population pushes the need for artificial fertilizers or greater incorporation of organic systems within agricultural infrastructure.

Currently, the best potential deep sea site, the Solwara 1 Project, has been found in the waters off Papua New Guinea, a high grade copper-gold resource and the world's first Seafloor Massive Sulphide (SMS) resource. The Solwara 1 Project is located at 1600 metres water depth in the Bismarck Sea, New Ireland Province. Using ROV (remotely operated underwater vehicles) technology, Nautilus Minerals Inc. is first company of its kind to announce plans to begin full-scale undersea excavation of mineral deposits. However a dispute with the government of Papua-New Guinea delayed production and its now scheduled to commence commercial operations in early 2018.

Laws and Regulations

The international law–based regulations on deep sea mining are contained in the United Nations Conventions on the Law of the Sea from 1973 to 1982, which came into force in 1994. The convention set up the International Seabed Authority (ISA), which regulates nations' deep sea mining ventures outside each nations' Exclusive Economic Zone (a 200-nautical-mile (370 km) area surrounding coastal nations). The ISA requires nations interested in mining to explore two equal mining sites and turn one over to the ISA, along with a transfer of mining technology over a 10- to 20-year period. This seemed reasonable at the time because it was widely believed that nodule mining would be extremely profitable. However, these strict requirements led some industrialized countries to refuse to sign the initial treaty in 1982.

Within the EEZ of nation states seabed mining comes under the jurisdiction of national laws. Despite extensive exploration both within and outside of EEZs, only a few countries, notably New Zealand, have established legal and institutional frameworks for the future development of deep seabed mining.

Papua New Guinea was the first country to approve a permit for the exploration of minerals in the deep seabed. Solwara 1 was awarded its licence and environmental permits despite three independent reviews of the environmental impact statement mine finding significant gaps and flaws in the underlying science.

The ISA has recently arranged a workshop in Australia where scientific experts, industry representatives, legal specialists and academics worked towards improving existing regulations and ensuring that development of seabed minerals does not cause serious and permanent damage to the marine environment.

Resources Mined

The deep sea contains many different resources available for extraction, including silver, gold, copper, manganese, cobalt, and zinc. These raw materials are found in various forms on the sea floor, usually in higher concentrations than terrestrial mines.

Minerals and Related Depths

Type of mineral deposit	Average Depth	Resources found
Polymetallic nodules	4,000 – 6,000 m	Nickel, copper, cobalt, and manganese
Manganese Crusts	800 – 2,400 m	Mainly cobalt, some vanadium, molybdenum and platinum
Sulfide deposits	1,400 – 3,700 m	Copper, lead and zinc some gold and silver

Diamonds are also mined from the seabed by De Beers and others. Nautilus Minerals Inc and Neptune Minerals are planning to mine the offshore waters of Papua New Guinea and New Zealand.

Extraction Methods

Recent technological advancements have given rise to the use remotely operated vehicles (ROVs) to collect mineral samples from prospective mine sites. Using drills and other cutting tools, the ROVs obtain samples to be analyzed for precious materials. Once a site has been located, a mining ship or station is set up to mine the area.

There are two predominant forms of mineral extraction being considered for full scale operations: continuous-line bucket system (CLB) and the hydraulic suction system. The CLB system is the preferred method of nodule collection. It operates much like a conveyor-belt, running from the sea floor to the surface of the ocean where a ship or mining platform extracts the desired minerals, and returns the tailings to the ocean. Hydraulic suction mining lowers a pipe to the seafloor which transfers nodules up to the mining ship. Another pipe from the ship to the seafloor returns the tailings to the area of the mining site.

In recent years, the most promising mining areas have been the Central and Eastern Manus Basin around Papua New Guinea and the crater of Conical Seamount to the east. These locations have shown promising amounts of gold in the area's sulfide deposits (an average of 26 parts per million). The relatively shallow water depth of 1050 m, along with the close proximity of a gold processing plant makes for an excellent mining site.

Environmental Impacts

Research shows that polymetallic nodule fields are hotspots of abundance and diversity for a highly vulnerable abyssal fauna. Because deep sea mining is a relatively new field, the complete consequences of full scale mining operations on this ecosystem are unknown. However, some researchers have said they believe that removal of parts of the sea floor will result in disturbances to the benthic layer, increased toxicity of the water column and sediment plumes from tailings. Removing parts of the sea floor could disturb the habitat of benthic organisms, with unknown long-term effects. Aside from the direct impact of mining the area, some researchers and environmental activists have raised concerns about leakage, spills and corrosion that could alter the mining area's chemical makeup.

Among the impacts of deep sea mining, sediment plumes could have the greatest impact. Plumes are caused when the tailings from mining (usually fine particles) are dumped back into the ocean, creating a cloud of particles floating in the water. Two types of plumes occur: near bottom plumes and surface plumes. Near bottom plumes occur when the tailings are pumped back down to the mining site. The floating particles increase the turbidity, or cloudiness, of the water, clogging filter-feeding apparatuses used by benthic organisms. Surface plumes cause a more serious problem. Depending on the size of the particles and water currents the plumes could spread over vast areas. The plumes could impact zooplankton and light penetration, in turn affecting the food web of the area.

Atmospheric Mining

Atmospheric mining is the process of extracting valuable materials or other non-renewable resources from the atmosphere. Due to the abundance of hydrogen and helium in the outer planets of the Solar System, atmospheric mining may be easier than mining terrestrial surfaces.

History of Atmospheric Mining

Atmospheric mining of outer planets has not yet begun.

Types of Atmospheric Mining

Hydrogen Mining

Hydrogen may fuel chemical and nuclear propulsion.

Helium Mining

Helium-3 may fuel nuclear propulsion.

Methane Mining

Methane may fuel chemical propulsion.

Exploration for Atmospheric Mining

Hydrogen and helium are abundant in outer planets.

Atmospheric composition of outer planets				
Resource	Jupiter	Saturn	Uranus	Neptune
Hydrogen	89.8	96.3	82.5	80.0
Helium	10.2	3.3	15.2	19.0
Methane			2.3	1.0
Other		0.4	1.0	

Methods of Atmospheric Mining

Aerostats

An aerostat would be a buoyant station in the atmosphere that gathers and stores gases. A vehicle would transfer the gases from the aerostat to an orbital station above the planet.

Scoopers

A scooper would be a vehicle that gathers and transfers gases from the atmosphere to an orbital station.

Cruisers

A cruiser would be a vehicle in the atmosphere that gathers and stores gases. A smaller vehicle would transfer the gases from the cruiser to an orbital station.

Biomining

Biomining is an approach to the extraction of desired minerals from ores. Microorganisms are used to leach out the minerals, rather than the traditional methods of extreme heat or toxic chemicals, which have a deleterious effect on the environment.

Overview

The development of industrial mineral processing has been established now in several countries including South Africa, Brazil and Australia. Iron-and sulfur-oxidizing microorganisms are used to release occluded copper, gold and uranium from mineral sulfides. Most industrial plants for biooxidation of gold-bearing concentrates have been operated at 40°C with mixed cultures of mesophilic bacteria of the genera *Acidithiobacillus* or *Leptospirillum ferrooxidans*. In subsequent studies the dissimulatory iron-reducing archaea *Pyrococcus furiosus* and *Pyrobaculum islandicum* were shown to reduce gold chloride to insoluble gold.

Using *Bacteria* such as *Acidithiobacillus ferrooxidans* to leach copper from mine tailings has improved recovery rates and reduced operating costs. Moreover, it permits extraction from low grade ores - an important consideration in the face of the depletion of high grade ores.

The potential applications of biotechnology to mining and processing are countless. Some examples of past projects in biotechnology include a biologically assisted in situ mining program, biodegradation methods, passive bioremediation of acid rock drainage, and bioleaching of ores and concentrates. This research often results in technology implementation for greater efficiency and productivity or novel solutions to complex problems. Additional capabilities include the bioleaching of metals from sulfide materials, phosphate ore bioprocessing, and the bioconcentration of metals from solutions. One project recently under investigation is the use of biological methods for the reduction of sulfur in coal-cleaning applications. From in situ mining to mineral processing and treatment technology, biotechnology provides innovative and cost-effective industry solutions.

The potential of thermophilic sulfide-oxidizing archaea in copper extraction has attracted interest due to the efficient extraction of metals from sulfide ores that are recalcitrant to dissolution. Microbial leaching is especially useful for copper ores because copper sulfate, as formed during the oxidation of copper sulfide ores, is very water-soluble. Approximately 25% of all copper mined worldwide is now obtained from leaching processes. The acidophilic archaea *Sulfolobus metallicus* and *Metallosphaera sedula* tolerate up to 4% of copper and have been exploited for mineral bio-mining. Between 40 and 60% copper extraction was achieved in primary reactors and more than 90% extraction in secondary reactors with overall residence times of about 6 days.

The oxidation of the ferrous ion (Fe^{2+}) to the ferric ion (Fe^{3+}) is an energy producing re-action for some microorganisms. As only a small amount of energy is obtained, large amounts of (Fe^{2+}) have to be oxidized. Furthermore, (Fe^{3+}) forms the insoluble $Fe(OH)_3$ precipitate in H_2O. Many Fe^{2+} oxidizing microorganisms also oxidize sulfur and are thus obligate acidophiles that further acidify the environment by the production of H_2SO_4. This is due in part to the fact that at neutral pH Fe^{2+} is rapidly oxidized chemically in contact with the air. In these conditions there is not enough Fe^{2+} to allow significant growth. At low pH, however, Fe^{2+} is much more stable. This explains why most of the Fe^{2+} oxidizing microorganisms are only found in acidic environments and are obligate acidophiles.

The best studied Fe^{2+} oxidizing bacterium is *Acidithiobacillus ferrooxidans*, an acidophililic chem-olithotroph. The microbiological oxidation of Fe^{2+} is an important aspect of the development of acidic pH's in mines, and constitutes a serious ecological problem. However, this process can also be usefully exploited when controlled. The sulfur containing ore pyrite (FeS_2) is at the start of this process. Pyrite is an insoluble crystalline structure that is abundant in coal- and mineral ores. It is produced by the following reaction:

$$S + FeS \rightarrow FeS_2$$

Normally pyrite is shielded from contact with oxygen and not accessible for microorganisms. Upon exploitation of the mine, however, pyrite is brought into contact with air (oxygen) and microorgan-isms and oxidation will start. This oxidation relies on a combination of chemically and microbio-logically catalyzed processes. Two electron acceptors can influence this process: O_2 and Fe^{3+} ions. The latter will only be present in significant amounts in acidic conditions (pH < 2.5). First a slow chemical process with O_2 as electron acceptor will initiate the oxidation of pyrite:

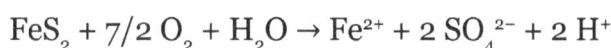

$$FeS_2 + 7/2\ O_2 + H_2O \rightarrow Fe^{2+} + 2\ SO_4^{2-} + 2\ H^+$$

This reaction acidifies the environment and the Fe^{2+} will be formed is rather stable. In such an environment *Acidithiobacillus ferrooxidans* will be able to grow rapidly. Upon further acidifica-tion *Ferroplasma* will also develop and further acidify. As a consequence of the microbial activity (energy producing reaction):

$$Fe^{2+} \rightarrow Fe^{3+}$$

This Fe^{3+} that remains soluble at low pH reacts spontaneously with the pyrite:

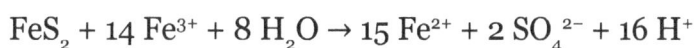

$$FeS_2 + 14\ Fe^{3+} + 8\ H_2O \rightarrow 15\ Fe^{2+} + 2\ SO_4^{2-} + 16\ H^+$$

The produced Fe^{2+} can again be used by the microorganisms and thus a cascade reaction will be initiated.

Processing Methods

In the industrial microbial leaching process popularly known as *bioleaching*, low grade ore is dumped in a large pile (the leach dump) and a dilute sulfuric acid solution (pH 2) is percolated down through the pile. The liquid coming out at the bottom of the pile, rich in the mineral is collected and transported to a precipitation plant where the metal is reprecipitated and purified. The liquid is then pumped back to the top of the pile and the cycle is repeated.

Acidithiobacillus ferrooxidans is able to oxidize the Cu^+ in chalcocite (Cu_2S) to Cu^{2+}, thus removing some of the copper in the soluble form, Cu^{2+}, and forming the mineral covellite (CuS). This oxidation of Cu^+ to Cu2+ is an energy yielding reaction (such as the oxidation of Fe^{2+} to Fe^{3+}). Covellite can then be oxidized, releasing sulfate and soluble Cu^{2+} as products.

A second mechanism, and probably the most important in most mining operations, involves chemical oxidation of the copper ore with ferric (Fe^{3+}) ions formed by the microbial oxidation of ferrous ions (derived from the oxidation of pyrite). Three possible reactions for the oxidation of copper ore are:

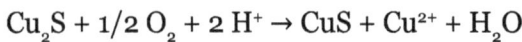

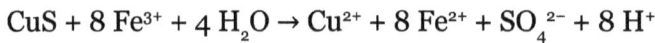

$$Cu_2S + 1/2\ O_2 + 2\ H^+ \rightarrow CuS + Cu^{2+} + H_2O$$

$$CuS + 2\ O_2 \rightarrow Cu^{2+} + SO_4^{2-}$$

$$CuS + 8\ Fe^{3+} + 4\ H_2O \rightarrow Cu^{2+} + 8\ Fe^{2+} + SO_4^{2-} + 8\ H^+$$

The copper metal is the recovered by using Fe^o from steel cans:

$$Fe^o + Cu^{2+} \rightarrow Cu^o + Fe^{2+}$$

The temperature inside the leach dump often rises spontaneously as a result of microbial activities. Thus, thermophilic iron-oxidizing chemolithotrophs such as thermophilic *Acidithiobacillus* species and *Leptospirillum* and at even higher temperatures the thermoacidophilic archaeon *Sulfolobus (Metallosphaera sedula)* may become important in the leaching process above 40 °C. Similarly to copper, *Acidithiobacillus ferrooxidans* can oxidize U^{4+} to U^{6+} with O_2 as electron acceptor. However, it is likely that the uranium leaching process depends more on the chemical oxidation of uranium by Fe^{3+}, with *At. ferrooxidans* contributing mainly through the reoxidation of Fe^{2+} to Fe^{3+} as described above.

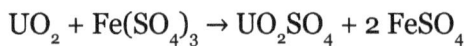

$$UO_2 + Fe(SO_4)_3 \rightarrow UO_2SO_4 + 2\ FeSO_4$$

Gold is frequently found in nature associated with minerals containing arsenic and pyrite. In the microbial leaching process *At. ferrooxidans* and relatives are able to attack and solubilize the arsenopyrite minerals, and in the process, releasing the trapped gold (Au):

$$2\ FeAsS[Au] + 7\ O_2 + 2\ H_2O + H_2SO_4 \rightarrow Fe(SO_4)_3 + 2\ H_3AsO_4 + [Au]$$

Biohydrometallurgy is an emerging trend in biomining in which commercial mining plants operate continuously stirred tank reactor (STR) and the airlift reactor (ALR) or pneumatic reactor (PR) of the Pachuca type to extract the low concentration mineral resources efficiently.

Brine Mining

Brine mining is the extraction of useful materials (elements or compounds) which are naturally dissolved in brine. The brine may be seawater, other surface water, or groundwater. It differs from solution mining or in-situ leaching in that those methods inject water or chemicals to dissolve materials which are in a solid state; in brine mining, the materials are already dissolved.

Brines are important sources of salt, iodine, lithium, magnesium, potassium, bromine, and other materials, and potentially important sources of a number of others.

Types of Brines Used for Mineral Extraction

Commercial brines include both surface water (seawater and saline lakes) and groundwater (shallow brine beneath saline or dry lakes, and deep brines in sedimentary basins). Brine brought to the surface by geothermal energy wells often contains high concentrations of minerals, but is not currently used for commercial mineral extraction. .

Seawater

Seawater has been used as a source of sea salt since prehistoric times, and more recently of magnesium and bromine. Potassium is sometimes recovered from the bittern left after salt precipitation. The oceans are often described as an inexhaustible resource.

Saline Lakes

There are many saline lakes with salinity greater than seawater, making them attractive for mineral extraction. Examples are the Dead Sea and the Great Salt Lake. In addition, some saline lakes, such as Lake Natron in East Africa, have chemistry very different than seawater, making them potential sources of sodium carbonate.

Shallow groundwater brines associated with saline or dry lakes.

The groundwater beneath saline or dry lakes often has brines with chemistry similar to that of the lakes or former lakes.

The chemistry of shallow brines used for mineral extraction is sometimes influenced by geothermal waters. This is true of a number of shallow brines in the western United States, such as at Searles Lake, California.

Geothermal Brines

Geothermal power plants often bring brine to the surface as part of the operation. This brine is usually re-injected into the ground, but some experiments have been made to extract minerals before re-injection. Brine brought to the surface by geothermal energy plants has been used in pilot plants as a source of colloidal silica (Wairkei, New Zealand, and Mammoth Lakes, California), and as a source of zinc (Salton Sea, California). Boron was recovered circa 1900 from geothermal steam at Larderello, Italy. Lithium recovery has also been investigated. But as of 2015, there is no sustained commercial-scale mineral recovery from geothermal brine.

Deep Brines in Sedimentary Basins

The concentration of dissolved solids in deep groundwater varies from much less than sea water to ten times the total dissolved solids of sea water. In general, total dissolved solids (TDS) concentrations increase with depth. Most deep groundwaters classified as brines (having total dissolved solids equal to or greater than that of seawater) are predominantly sodium chloride type. However, the predominance of chloride usually increases with increasing TDS, at the expense of sulfate. The ratio of calcium to sodium usually increases with depth.

The presence of groundwater with TDS higher than seawater is in some cases due to contact with salt beds. More often, however, the higher TDS of deep sediments is thought to be the result of the sediments acting as semi-permeable membranes. As the sediments compact under burial pressure, the dissolved species are less mobile than the water, resulting in higher TDS concentrations than seawater. Bivalent species such as calcium (Ca^{+2}) are less mobile than univalent species such as sodium (Na^{+}), resulting in calcium enrichment. The ratio of potassium to sodium (K/Na) may increase or decrease with depth, thought to be the result of ion exchange with the sediments.

Materials Recovered from Brines

Many brines contain more than one recovered product. For instance, the shallow brine beneath Searles Lake, California, is or has been a source of borax, potash, bromine, lithium, phosphate, soda ash, and sodium sulfate.

Salt

Source	Salt concentration
Seawater	129,500 mg/l

Salt (sodium chloride) has been a valuable commodity since prehistoric times, and its extraction from sea water also goes back to prehistory. Salt is extracted from seawater in many countries around the world, but the majority of salt put on the market today is mined from solid evaporite deposits.

Salt is produced as a byproduct of potash extraction from Dead Sea brine at one plant in Israel (Dead Sea Works), and another in Jordan (Arab Salt Works). The total salt precipitated in solar evaporation at the Dead Sea plants is tens of millions of tons annually, but very little of the salt is marketed.

Today, salt from groundwater brines is generally a byproduct of the process of extracting other dissolved substances from brines and constitutes only a small part of world salt production. In the United States, salt is recovered from surface brine at the Great Salt Lake, Utah, and from a shallow subsurface brine at Searles Lake, California.

Sodium Sulfate

In 1997 about two-thirds of world sodium sulfate production was recovered from brine. Two plants in the US, at Searles Lake, California, and Seagraves, Texas, recovered sodium sulfate from shallow brines beneath dry lakes.

Soda Ash

Soda ash (sodium carbonate) is recovered from shallow subsurface brines at Searles Lake, California.

Colloidal Silica

Brines brought to the surface by geothermal energy production often contain concentrations of dissolved silica of about 500 parts per million. A number of geothermal plants have pilot-tested recovery of colloidal silica, including those at Wairakei, New Zealand, Mammoth Lakes, California, and the Salton Sea, California. To date, colloidal silica from brine has not achieved commercial production.

Potash

Location	Potassium concentration	Source
Ocean	380 mg/l	Seawater
Ocean	17,700 mg/l	Seawater, bittern remaining after salt precipitation
Salar de Olaroz mine, Argentina	5,730 mg/l	Shallow brine beneath dry lake
Salar de Atacama, Chile	19,400 mg/l	Shallow brine beneath dry lake
Da Chaidam Salt Lake, China	22,500 mg/l	Saline lake
Dead Sea, Israel and Jordan	6,200 mg/l	Saline lake

Potash is recovered from surface brine of the Dead Sea, at plants in Israel and Jordan. In 2013 Dead Sea brine provided 9.2% of the world production of potash. As of 1996, the Dead Sea was estimated to contain 2.05 million tons of potassium chloride, the largest brine reserve of potassium other than the ocean.

Lithium

Location	Lithium concentration	Source
Ocean	0.17 mg/l	Seawater
Clayton Valley, Nevada	300 mg/l	Shallow brine beneath dry lake
Salton Sea, California	270 mg/l	Geothermal brine
Salar de Olaroz mine, Argentina	690 mg/l	Shallow brine beneath dry lake

In 2015 subsurface brines yielded about half of the world's lithium production. Whereas seawater contains about 0.17 mg/l, subsurface brines may contain up to 4,000 mg/l, more than four orders of magnitude greater than sea water. Typical commercial lithium concentrations are between 200 and 1,400 mg/l.

The largest operations are in the shallow brine beneath the Salar de Atacama dry lakebed in Chile, which as of 2015 yielded about a third of the world's supply. The brine operations are primarily for potassium; extraction of lithium as a byproduct began in 1997.

The shallow brine beneath the Salar de Uyuni in Bolivia is thought to contain the world's largest lithium resource, often estimated to be half or more of the world's resource. As of 2015, no commercial extraction has taken place, other than a pilot plant.

Commercial deposits of shallow lithium brines beneath dry lakebeds have the following characteristics in common:

- Arid climate

- Closed basin with a dry or seasonal lake

- Tectonically-driven subsidence

- Igneous or geothermal activity

- Lithium-rich source rock

- Permeable aquifers

- Enough time to concentrate brine

In 2010 Simbol Materials received a $3 million grant from the U.S. Department of Energy for a pilot project aimed at showing the financial feasibility of extracting high-quality lithium from geothermal brine. It uses brine from the 49.9 megawatt Featherstone geothermal power plant in California's Imperial Valley. Simbol passes the plant's extracted fluid through a series of membranes, filters and adsorption materials to extract lithium.

Boron

Location	Boron concentration	Source
Ocean	4.6 mg/l	Seawater
Salar de Olaroz, Argentina	1,050 mg/l	Shallow brine beneath dry lake

Boron is recovered from shallow brines beneath Searles Lake, California, by Searles Valley Minerals. Although boron is the primary product, potassium and other salts are also recovered as byproducts.

The brine beneath the Salar de Olaroz, Argentina, is a commercial source of boron, lithium, and potassium.

Circa 1900, boron was recovered from geothermal steam at Larderello, Italy.

Iodine

Location	Iodine concentration	Source
Ocean	0.06 mg/l	Seawater
Kanto Gas Field, Japan	160 mg/l	Deep brine in sedimentary basin
Morrow Sandstone, Oklahoma, USA	300 mg/l	Deep brine in sedimentary basin

Brines are a major source of iodine supply worldwide. Major deposits occur in Japan and the United States. Iodine is recovered from deep brines pumped to the surface as a bypoduct of oil and natural gas production. Seawater contains about 0.06 mg/l iodine, while subsurface brines contain as much as 1,560 mg/l, more than five orders of magnitude greater than seawater. The source of the iodine is thought to be organic material in shales, which also form the source rock for the associated hydrocarbons.

Japan

By far the largest source of iodine from brine is Japan, where iodine-rich water is co-produced with natural gas. Iodine extraction began in 1934. In 2013 seven companies were reported to be extracting iodine. Japanese iodine brines are produced from mostly marine sediments ranging in age from Pliocene to Pleistocene. The main producing area is the Southern Kanto gas field on the east-central coast of Honshu. The iodine content of the brine can be as high as 160 ppm.

Anadarko Basin, Oklahoma

Since 1977, iodine has been extracted from brine in the Morrow Sandstone of Pennsylvanian age, at locations in the Anadarko Basin. of northwest Oklahoma. The brine occurs at depths of 6,000 to 10,000 feet, and contains about 300 ppm iodine.

Bromine

Location	Bromine concentration	Source
Ocean	65 mg/l	Seawater
Ocean	2,970 mg/l	Seawater, bittern remaining after salt precipitation
Smackover Formation, Arkansas, USA	5,000 to 6,000 mg/l	Deep brine in sedimentary basin
Dead Sea, Israel and Jordan	10,000 mg/l	Saline lake

All the world's bromine production is derived from brine. The majority is recovered from Dead Sea brine at plants in Israel and Jordan, where bromine is a byproduct of potash recovery. Plants in the United States, China, Turkmenistan, and Ukraine, recover bromine from subsurface brines. In India and Japan, bromine is recovered as a byproduct of sea salt production.

Magnesium and Magnesium Compounds

Location	Magnesium concentration	Source
Ocean	1,350 mg/l	Seawater
Ocean	56,100 mg/l	Seawater, bittern remaining after salt precipitation
Dead Sea, Israel and Jordan	35,200 mg/l	Saline lake

The first commercial production of magnesium from seawater was recorded in 1923, when some solar salt plants around San Francisco Bay, California, extracted magnesium from the bitterns left after salt precipitation.

The Dow Chemical Company began producing magnesium on a small scale in 1916, from deep subsurface brine in the Michigan Basin. In 1933, Dow began using an ion exchange process to concentrate the magnesium in its brine. In 1941, prompted by the need for magnesium for aircraft during World War II, Dow started a large plant at Freeport, Texas, to extract magnesium from the sea. A number of other plants to extract magnesium from brine were built in the US, including one near the Freeport plant at Velasco. At the end of World War II, all shut down except the plant at

Freeport, Texas, although the Velasco plant was reactivated during the Korean War. The magnesium plant at Freeport operated until 1998, when Dow announced that it would not rebuild the unit following hurricane damage.

Because metallic magnesium is extracted from brine by an electrolytic process, the economics are sensitive to the cost of electricity. Dow had located their facility on the Texas coast to take advantage of cheap natural gas for electrical generation. In 1951, Norsk Hydro started a magnesium-from-seawater plant at Heroya, Norway, supplied by inexpensive hydroelectricity. The two seawater magnesium plants, in Texas and Norway, provided more than half the world's primary magnesium through the 1950s and 1960s.

As of 2014, the only producer of primary magnesium metal in the United States was U.S. Magnesium LLC, which extracted the metal from surface brine of the Great Salt Lake, at its plant in Rowley, Utah.

The Dead Sea Works in Israel produces magnesium as a byproduct of potash extraction.

Zinc

Location	Zinc concentration	Source
Ocean	0.01 mg/l	Seawater
Salton Sea, California	270 mg/l	Geothermal brine

Starting in 2002, CalEnergy extracted zinc from brines at its geothermal energy plants at the Salton Sea, California. At full production, the company hoped to produce 30,000 metric tons of 99.99% pure zinc per year, yielding about as much profit as the company made from geothermal energy. But the zinc recovery unit did not perform as anticipated, and zinc recovery halted in 2004.

Tungsten

Location	Tungsten concentration	Source
Ocean	0.0001 mg/l	Seawater
Searles Lake, California	56 mg/l	Shallow brine beneath dry lake

Some near-surface brines in the western United States contain anomalously high concentrations of dissolved tungsten. Should recovery ever prove economic, some brines could be significant sources of tungsten. For instance, brines beneath Searles Lake, California, with concentrations of about 56 mg/l tungsten (70 mg/l WO_3), contain about 8.5 million short tons of tungsten. Although 90% of the dissolved tungsten is technically recoverable by ion exchange resins, recovery is uneconomic.

Uranium

Source	Uranium concentration
Seawater	0.003 mg/l

In 2012 research for the US Department of Energy, building on Japanese research from the 1990s, tested a method for extracting uranium from seawater, which, they concluded, could extract uranium at a cost of US$660/kg. While this was still five times the cost of uranium from ore, the amount of uranium dissolved in seawater would be enough to provide nuclear fuel for thousands of years at current rates of consumption.

Gold

Source	Gold concentration
Seawater	0.000004 mg/l

Attempts to extract gold from seawater were common in the early 20th century. A number of people claimed to be able to economically recover gold from sea water, but they were all either mistaken or acted in an intentional deception. Prescott Jernegan ran a gold-from-seawater swindle in the United States in the 1890s. A British fraudster ran the same scam in England in the early 1900s.

Fritz Haber (the German inventor of the Haber process) did research on the extraction of gold from sea water in an effort to help pay Germany's reparations following World War I. Based on published values of 2 to 64 ppb of gold in seawater, a commercially successful extraction seemed possible. After analysis of 4,000 water samples yielding an average of 0.004 ppb, it became clear to Haber that the extraction would not be possible, and he stopped the project.

Uranium Mining

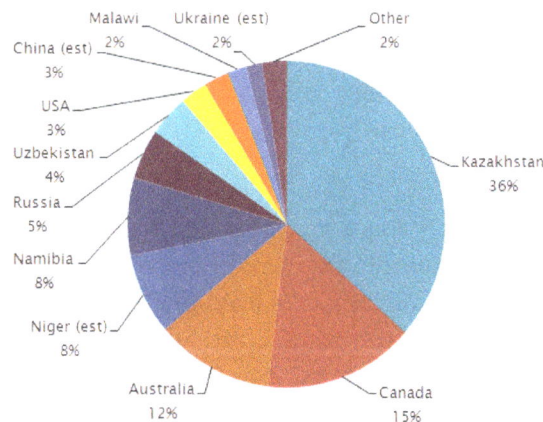

World Uranium Mining Production 2012

2012 uranium mining, by nationality.

Uranium mining is the process of extraction of uranium ore from the ground. The worldwide production of uranium in 2012 amounted to 58,394 tonnes. Kazakhstan, Canada, and Australia are the top three producers and together account for 64% of world uranium production. Other important uranium producing countries in excess of 1,000 tons per year are Niger, Namibia, Russia, Uzbekistan, the United States, China, and Malawi.

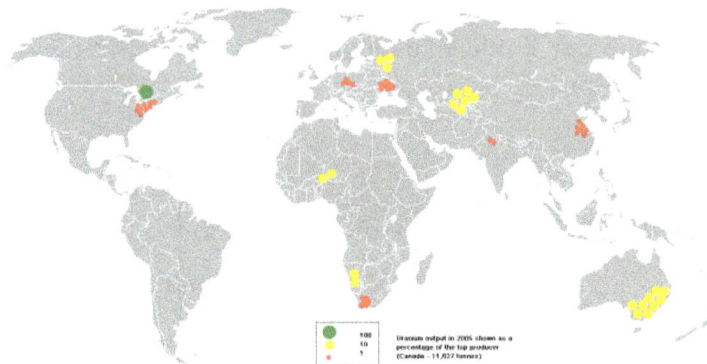

World Uranium production in 2005.

Uranium ores are normally processed by grinding the ore materials to a uniform particle size and then treating the ore to extract the uranium by chemical leaching. The milling process commonly yields dry powder-form material consisting of natural uranium, "yellowcake," which is sold on the uranium market as U_3O_8.

Uranium from mining is used almost entirely as fuel for nuclear power plants. As of July 2014, the price of uranium concentrate remained near a five-year low, the uranium price having fallen more than 50% from the peak spot price in January 2011, reflecting the loss of Japanese demand following the 2011 Fukushima nuclear disaster. As a result of continued low prices, in February 2014 mining company Cameco deferred plans to expand production from existing Canadian mines, although it continued work to open a new mine at Cigar Lake. Also in February 2014, Paladin energy suspended operations at its mine in Malawi, saying that the high-cost operation was losing money at current prices.

History

Uranium minerals were noticed by miners for a long time prior to the discovery of uranium in 1789. The uranium mineral pitchblende, also known as uraninite, was reported from the Krušné hory (Ore Mountains), Saxony, as early as 1565. Other early reports of pitchblende date from 1727 in Jáchymov and 1763 in Schwarzwald.

In the early 19th century, uranium ore was recovered as a byproduct of mining in Saxony, Bohemia, and Cornwall. The first deliberate mining of radioactive ores took place in Jáchymov, a silver-mining city in the Czech Republic. Marie Curie used pitchblende ore from Jáchymov to isolate the element radium, a decay product of uranium. Until World War II uranium mining was done primarily for the radium content. Sources for radium, contained in the uranium ore, were sought for use as luminous paint for watch dials and other instruments, as well as for health-related applications, some of which in retrospect were certainly harmful. The byproduct uranium was used mostly as a yellow pigment.

In the United States, the first radium/uranium ore was discovered in 1871 in gold mines near Central City, Colorado. This district produced about 50 tons of high grade ore between 1871 and 1895. However, most American uranium ore before World War II came from vanadium deposits on the Colorado Plateau of Utah and Colorado.

In Cornwall, the South Terras Mine near St. Stephen opened for uranium production in 1873, and produced about 175 tons of ore before 1900. Other early uranium mining occurred in Autunois in France's Massif Central, Oberpfalz in Bavaria, and Billingen in Sweden.

The Shinkolobwe deposit in Katanga, Belgian Congo now Shaba Province, Democratic Republic of the Congo (DRC) was discovered in 1913, and exploited by the Union Minière du Haut Katanga. Other important early deposits include Port Radium, near Great Bear Lake, Canada discovered in 1931, along with Beira Province, Portugal; Tyuya Muyun, Uzbekistan, and Radium Hill, Australia.

Because of the need for the uranium for bomb research during World War II, the Manhattan Project used a variety of sources for the element. The Manhattan Project initially purchased uranium ore from the Belgian Congo, through the Union Minière du Haut Katanga. Later the project contracted with vanadium mining companies in the American Southwest. Purchases were also made from the Eldorado Mining and Refining Limited company in Canada. This company had large stocks of uranium as waste from its radium refining activities.

American uranium ores mined in Colorado were mixed ores of vanadium and uranium, but because of wartime secrecy, the Manhattan Project would publicly admit only to purchasing the vanadium, and did not pay the uranium miners for the uranium content. In a much later lawsuit, many miners were able to reclaim lost profits from the U.S. government. American ores had much lower uranium concentrations than the ore from the Belgian Congo, but they were pursued vigorously to ensure nuclear self-sufficiency.

Similar efforts were undertaken in the Soviet Union, which did not have native stocks of uranium when it started developing its own atomic weapons program.

Intensive exploration for uranium started after the end of World War II as a result of the military and civilian demand for uranium. There were three separate periods of uranium exploration or "booms." These were from 1956 to 1960, 1967 to 1971, and from 1976 to 1982.

In the 20th century the United States was the world's largest uranium producer. Grants Uranium District in northwestern New Mexico was the largest United States uranium producer. The Gas Hills Uranium District, was the second largest uranium producer. The famous Lucky Mc Mine is located in the Gas Hills near Riverton, Wyoming. Canada has since surpassed the United States as the cumulative largest producer in the world. In 1990, 55% of world production came from underground mines, but this shrunk dramatically to 1999, with 33% then. From 2000 the new Canadian mines increased it again, and with Olympic Dam it is now 37%. In situ leach (ISL, or ISR) mining has been steadily increasing its share of the total, mainly due to Kazakhstan.

Types of Uranium Deposits

Many different types of uranium deposits have been discovered and mined. There are mainly three types of uranium deposits including unconformity-type deposits, namely paleoplacer deposits and sandstone-type also known as roll front type deposits.

Uranium deposits are classified into 15 categories according to their geological setting and the type of rock in which they are found. This geological classification system is determined by the International Atomic Energy Agency (IAEA).

Uranium Deposits in Sedimentary Rock

The Mi Vida uranium mine, near Moab, Utah. Note alternating red and white/green sandstone.
This type of uranium deposit is easier and cheaper to mine than the other types because the uranium
is found not far from the surface of the Crust.

Uranium deposits in sedimentary rocks include those in sandstone (in Canada and the western US), Precambrian unconformities (in Canada), phosphate, Precambrian quartz-pebble conglomerate, collapse breccia pipes, and calcrete.

Sandstone uranium deposits are generally of two types. Roll-front type deposits occur at the boundary between the up dip and oxidized part of a sandstone body and the deeper down dip reduced part of a sandstone body. Peneconcordant sandstone uranium deposits, also called Colorado Plateau-type deposits, most often occur within generally oxidized sandstone bodies, often in localized reduced zones, such as in association with carbonized wood in the sandstone.

Precambrian quartz-pebble conglomerate-type uranium deposits occur only in rocks older than two billion years old. The conglomerates also contain pyrite. These deposits have been mined in the Blind River-Elliot Lake district of Ontario, Canada, and from the gold-bearing Witwatersrand conglomerates of South Africa.

Unconformity-type deposits make up about 33% of the World Outside Centrally Planned Economies Areas (WOCA)'s uranium deposits.

Igneous or Hydrothermal Uranium Deposits

Hydrothermal uranium deposits encompass the vein-type uranium ores. Igneous deposits include nepheline syenite intrusives at Ilimaussaq, Greenland; the disseminated uranium deposit at Rossing, Namibia; and uranium-bearing pegmatites. Disseminated deposits are also found in the states of Washington and Alaska in the US.

Breccia Uranium Deposits

Breccia uranium deposits are found in rocks that have been broken due to tectonic fracturing, or weathering. Breccia uranium deposits are most common in India, Australia and the United States.

Olympic Dam mine is the world's largest uranium deposit and home to the Olympic Dam Centre, a mining company currently owned by BHP Billiton.

Exploration

Uranium prospecting is similar to other forms of mineral exploration with the exception of some specialized instruments for detecting the presence of radioactive isotopes.

The Geiger counter was the original radiation detector, recording the total count rate from all energy levels of radiation. Ionization chambers and Geiger counters were first adapted for field use in the 1930s. The first transportable Geiger–Müller counter (weighing 25 kg) was constructed at the University of British Columbia in 1932. H.V. Ellsworth of the GSC built a lighter weight, more practical unit in 1934. Subsequent models were the principal instruments used for uranium prospecting for many years, until geiger counters were replaced by scintillation counters.

The use of airborne detectors to prospect for radioactive minerals was first proposed by G.C. Ridland, a geophysicist working at Port Radium in 1943. In 1947, the earliest recorded trial of airborne radiation detectors (ionization chambers and Geiger counters) was conducted by Eldorado Mining and Refining Limited. (a Canadian Crown Corporation since sold to become Cameco Corporation). The first patent for a portable gamma-ray spectrometer was filed by Professors Pringle, Roulston & Brownell of the University of Manitoba in 1949, the same year as they tested the first portable scintillation counter on the ground and in the air in northern Saskatchewan.

Airborne gamma-ray spectrometry is now the accepted leading technique for uranium prospecting with worldwide applications for geological mapping, mineral exploration & environmental monitoring. Airborne gamma-ray spectrometry used specifically for uranium measurement and prospecting must account for a number of factors like the distance between the source and the detector and the scattering of radiation through the minerals, surrounding earth and even in the air. In Australia, a Weathering Intensity Index has been developed to help prospectors based on the Shuttle Radar Topography Mission (SRTM) elevation and airborne gamma-ray spectrometry images.

A deposit of uranium, discovered by geophysical techniques, is evaluated and sampled to determine the amounts of uranium materials that are extractable at specified costs from the deposit. Uranium reserves are the amounts of ore that are estimated to be recoverable at stated costs.

Mining Techniques

As with other types of hard rock mining there are several methods of extraction. In 2012, the percentage of the mined uranium produced by each mining method was: in-situ leach (44.9 percent), underground mining (26.2 percent), open pit (19.9 percent), and heap leaching (1.7 percent). The remaining 7.3% was derived as a byproduct of mining for other minerals, and miscellaneous recovery.

Open Pit

In open pit mining, overburden is removed by drilling and blasting to expose the ore body, which is then mined by blasting and excavation using loaders and dump trucks. Workers spend much time in enclosed cabins thus limiting exposure to radiation. Water is extensively used to suppress airborne dust levels.

Rössing open pit uranium mine, Namibia

Underground Uranium Mining

If the uranium is too far below the surface for open pit mining, an underground mine might be used with tunnels and shafts dug to access and remove uranium ore. There is less waste material removed from underground mines than open pit mines, however this type of mining exposes underground workers to the highest levels of radon gas.

Underground uranium mining is in principle no different from any other hard rock mining and other ores are often mined in association (e.g., copper, gold, silver). Once the ore body has been identified a shaft is sunk in the vicinity of the ore veins, and crosscuts are driven horizontally to the veins at various levels, usually every 100 to 150 metres. Similar tunnels, known as drifts, are driven along the ore veins from the crosscut. To extract the ore, the next step is to drive tunnels, known as raises when driven upwards and winzes when driven downwards through the deposit from level to level. Raises are subsequently used to develop the stopes where the ore is mined from the veins.

The stope, which is the workshop of the mine, is the excavation from which the ore is extracted. Two methods of stope mining are commonly used. In the "cut and fill" or open stoping method, the space remaining following removal of ore after blasting is filled with waste rock and cement. In the "shrinkage" method, only sufficient broken ore is removed via the chutes below to allow miners working from the top of the pile to drill and blast the next layer to be broken off, eventually leaving a large hole. Another method, known as room and pillar, is used for thinner, flatter ore bodies. In this method the ore body is first divided into blocks by intersecting drives, removing ore while so doing, and then systematically removing the blocks, leaving enough ore for roof support.

The health effects discovered from radon exposure in unventilated uranium mining prompted the switch away from uranium mining via tunnel underground mining towards open cut and In-situ leaching technology, a method of extraction that does not produce the same occupational hazards, or mine tailings, as conventional mining.

With regulations in place to ensure the use of high volume ventilation technology if any confined space uranium mining is occurring, with both largely eliminating occupational exposure and mining deaths. The Olympic Dam and Canadian underground mines are ventilated with powerful fans with Radon levels being kept at a very low to practically "safe level" in uranium mines. Naturally occurring Radon in other, non-uranium mines, also may need control by ventilation.

Heap Leaching

Heap leaching is an extraction process by which chemicals (usually sulfuric acid) are used to extract the economic element from ore which has been mined and placed in piles on the surface. Heap leaching is generally economically feasible only for oxide ore deposits. Oxidation of sulfide deposits occurs during the geological process called weathering. Therefore, oxide ore deposits are typically found close to the surface. If there are no other economic elements within the ore a mine might choose to extract the uranium using a leaching agent, usually a low molar sulfuric acid.

If the economic and geological conditions are right, the mining company will level large areas of land with a small gradient, layering it with thick plastic (usually HDPE or LLDPE), sometimes with clay, silt or sand beneath the plastic liner. The extracted ore will typically be run through a crusher and placed in heaps atop the plastic. The leaching agent will then be sprayed on the ore for 30–90 days. As the leaching agent filters through the heap the uranium will break its bonds with the oxide rock and enter the solution. The solution will then filter along the gradient into collecting pools which will then be pumped to on-site plants for further processing. Only some of the uranium (commonly about 70%) is actually extracted.

The uranium concentrations within the solution are very important for the efficient separation of pure uranium from the acid. As different heaps will yield different concentrations the solution is pumped to a mixing plant that is carefully monitored. The properly balanced solution is then pumped into a processing plant where the Uranium is separated from the sulfuric acid.

Heap leach is significantly cheaper than traditional milling processes. The low costs allow for lower grade ore to be economically feasible (given that it is the right type of ore body). Environmental law requires that the surrounding ground water is continually monitored for possible contamination. The mine will also have to have continued monitoring even after the shutdown of the mine. In the past mining companies would sometimes go bankrupt, leaving the responsibility of mine reclamation to the public. Recent additions to the mining law require that companies set aside the money for reclamation before the beginning of the project. The money will be held by the public to insure adherence to environmental standards if the company were to ever go bankrupt.

Another very similar mining technique is called in situ, or in place mining where the ore doesn't even need extracting.

In-situ Leaching

In-situ leaching (ISL), also known as solution mining, or in-situ recovery (ISR) in North America, involves leaving the ore where it is in the ground, and recovering the minerals from it by dissolving them and pumping the pregnant solution to the surface where the minerals can be recovered. Consequently, there is little surface disturbance and no tailings or waste rock generated. However, the orebody needs to be permeable to the liquids used, and located so that they do not contaminate ground water away from the orebody.

Trial well field for in-situ recovery at Honeymoon, South Australia

Uranium ISL uses the native groundwater in the orebody which is fortified with a complexing agent and in most cases an oxidant. It is then pumped through the underground orebody to recover the minerals in it by leaching. Once the pregnant solution is returned to the surface, the uranium is recovered in much the same way as in any other uranium plant (mill).

In Australian ISL mines (Beverley, Four Mile and Honeymoon Mine) the oxidant used is hydrogen peroxide and the complexing agent sulfuric acid. Kazakh ISL mines generally do not employ an oxidant but use much higher acid concentrations in the circulating solutions. ISL mines in the USA use an alkali leach due to the presence of significant quantities of acid-consuming minerals such as gypsum and limestone in the host aquifers. Any more than a few percent carbonate minerals means that alkali leach must be used in preference to the more efficient acid leach.

The Australian government has published a best practice guide for in situ leach mining of uranium, which is being revised to take account of international differences.

Recovery from Seawater

The uranium concentration of sea water is low, approximately 3.3 parts per billion or 3.3 micrograms per liter of seawater. But the quantity of this resource is gigantic and some scientists believe this resource is practically limitless with respect to world-wide demand. That is to say, if even a portion of the uranium in seawater could be used the entire world's nuclear power generation fuel could be provided over a long time period. Some anti-nuclear proponents claim this statistic is exaggerated. Although research and development for recovery of this low-concentration element by inorganic adsorbents such as titanium oxide compounds has occurred since the 1960s in the United Kingdom, France, Germany, and Japan, this research was halted due to low recovery efficiency.

At the Takasaki Radiation Chemistry Research Establishment of the Japan Atomic Energy Research Institute (JAERI Takasaki Research Establishment), research and development has continued culminating in the production of adsorbent by irradiation of polymer fiber. Adsorbents have been synthesized that have a functional group (amidoxime group) that selectively adsorbs heavy metals, and the performance of such adsorbents has been improved. Uranium adsorption capacity of the polymer fiber adsorbent is high, approximately tenfold greater in comparison to the conventional titanium oxide adsorbent.

One method of extracting uranium from seawater is using a uranium-specific nonwoven fabric as an adsorbent. The total amount of uranium recovered from three collection boxes containing 350 kg of fabric was >1 kg of yellowcake after 240 days of submersion in the ocean. According to the OECD, uranium may be extracted from seawater using this method for about $300/kg-U. The experiment by Seko *et al.* was repeated by Tamada et al. in 2006. They found that the cost varied from ¥15,000 to ¥88,000 depending on assumptions and "The lowest cost attainable now is ¥25,000 with 4g-U/kg-adsorbent used in the sea area of Okinawa, with 18 repetitionuses [*sic*]." With the May, 2008 exchange rate, this was about $240/kg-U.

In 2012, ORNL researchers announced the successful development of a new adsorbent material dubbed "HiCap", which vastly outperforms previous best adsorbents, which perform surface retention of solid or gas molecules, atoms or ions. "We have shown that our adsorbents can extract five to seven times more uranium at uptake rates seven times faster than the world's best adsorbents," said Chris Janke, one of the inventors and a member of ORNL's Materials Science and Technology Division. HiCap also effectively removes toxic metals from water, according to results verified by researchers at Pacific Northwest National Laboratory.

Uranium Prices

Since 1981 uranium prices and quantities in the US are reported by the Department of Energy. The import price dropped from 32.90 US$/lb-$U_3O_8$ in 1981 down to 12.55 in 1990 and to below 10 US$/lb-$U_3O_8$ in the year 2000. Prices paid for uranium during the 1970s were higher, 43 US$/lb-$U_3O_8$ is reported as the selling price for Australian uranium in 1978 by the Nuclear Information Centre. Uranium prices reached an all-time low in 2001, costing US$7/lb, but in April 2007 the price of Uranium on the spot market rose to US$113.00/lb, a high point of the uranium bubble of 2007. This was very close to the all time high (adjusted for inflation) in 1977.

Following the 2011 Fukushima nuclear disaster, the global uranium sector remains depressed with the uranium price falling more than 50%, declining share values, and reduced profitability of uranium producers since March 2011 and into 2014. As a result, uranium companies worldwide are reducing costs, and limiting operations.

Politics of Uranium Mining

In the beginning of the Cold War, to ensure adequate supplies of uranium for national defense, the United States Congress passed the U.S. Atomic Energy Act of 1946, creating the Atomic Energy Commission (AEC) which had the power to withdraw prospective uranium mining land from public purchase, and also to manipulate the price of uranium to meet national needs. By setting a high price for uranium ore, the AEC created a uranium "boom" in the early 1950s, which attracted many prospectors to the Four Corners region of the country. Moab, Utah became known as the Uranium-capital of the world, when geologist Charles Steen discovered such an ore in 1952, even though American ore sources were considerably less potent than those in the Belgian Congo or South Africa.

In the 1950s methods for extracting diluted uranium and thorium, found in abundance in granite or seawater, were pursued. Scientists speculated that, used in a breeder reactor, these materials would potentially provide limitless source of energy.

American military requirements declined in the 1960s, and the government completed its uranium procurement program by the end of 1970. Simultaneously, a new market emerged: commercial nuclear power plants. However, in the U.S. this market virtually collapsed by the end of the 1970s as a result of industrial strains caused by the energy crisis, popular opposition, and finally the Three Mile Island nuclear accident in 1979, all of which led to a *de facto* moratorium on the development of new nuclear reactor power stations.

In Europe a mixed situation exists. Considerable nuclear power capacities have been developed, notably in Belgium, Finland. France, Germany, Spain, Sweden, Switzerland and the UK. In many countries development of nuclear power has been stopped and phased out by legal actions. In Italy the use of nuclear power was barred by a referendum in 1987, however this is now under revision. Ireland in 2008 also had no plans to change its non-nuclear stance, although since the opening in 2012 of the East-West Interconnector between Ireland and Britain, it has been supported by British nuclear power.

The years 1976 and 1977 saw uranium mining become a major political issue in Australia, with the Ranger Inquiry (Fox) report opening up a public debate about uranium mining. The Movement Against Uranium Mining group was formed in 1976, and many protests and demonstrations against uranium mining were held. Concerns relate to the health risks and environmental damage from uranium mining. Notable Australian anti-uranium activists have included Kevin Buzzacott, Jacqui Katona, Yvonne Margarula, and Jillian Marsh.

The World Uranium Hearing was held in Salzburg, Austria in September 1992. Anti-nuclear speakers from all continents, including indigenous speakers and scientists, testified to the health and environmental problems of uranium mining and processing, nuclear power, nuclear weapons, nuclear tests, and radioactive waste disposal. People who spoke at the 1992 Hearing include: Thomas Banyacya, Katsumi Furitsu, Manuel Pino and Floyd Red Crow Westerman. They highlighted the threat of radioactive contamination to all peoples, especially indigenous communities and said that their survival requires self-determination and emphasis on spiritual and cultural values. Increased renewable energy commercialization was advocated.

Health Risks of Uranium Mining

Lung Cancer Deaths

Uranium ore emits radon gas. The health effects of high exposure to radon are a particular problem in the mining of uranium; significant excess lung cancer deaths have been identified in epidemiological studies of uranium miners employed in the 1940s and 1950s.

The first major studies with radon and health occurred in the context of uranium mining, first in the Joachimsthal region of Bohemia and then in the Southwestern United States during the early Cold War. Because radon is a product of the radioactive decay of uranium, underground uranium mines may have high concentrations of radon. Many uranium miners in the Four Corners region contracted lung cancer and other pathologies as a result of high levels of exposure to radon in the mid-1950s. The increased incidence of lung cancer was particularly pronounced among Native American and Mormon miners, because those groups normally have low rates of lung cancer. Safety standards requiring expensive ventilation were not widely implemented or policed during this period.

In studies of uranium miners, workers exposed to radon levels of 50 to 150 picocuries of radon per liter of air (2000–6000 Bq/m^3) for about 10 years have shown an increased frequency of lung cancer. Statistically significant excesses in lung cancer deaths were present after cumulative exposures of less than 50 WLM. There is, however, unexplained heterogeneity in these results (whose confidence interval do not always overlap). The size of the radon-related increase in lung cancer risk varied by more than an order of magnitude between the different studies.

Since that time, ventilation and other measures have been used to reduce radon levels in most affected mines that continue to operate. In recent years, the average annual exposure of uranium miners has fallen to levels similar to the concentrations inhaled in some homes. This has reduced the risk of occupationally induced cancer from radon, although it still remains an issue both for those who are currently employed in affected mines and for those who have been employed in the past. The power to detect any excess risks in miners nowadays is likely to be small, exposures being much smaller than in the early years of mining.

Clean-up Efforts

United States

Despite efforts made in cleaning up uranium sites, significant problems stemming from the legacy of uranium development still exist today on the Navajo Nation and in the states of Utah, Colorado, New Mexico, and Arizona. Hundreds of abandoned mines have not been cleaned up and present environmental and health risks in many communities. At the request of the U.S. House Committee on Oversight and Government Reform in October 2007, and in consultation with the Navajo Nation, the Environmental Protection Agency (EPA), along with the Bureau of Indian Affairs (BIA), the Nuclear Regulatory Commission (NRC), the Department of Energy (DOE), and the Indian Health Service (IHS), developed a coordinated Five-Year Plan to address uranium contamination. Similar interagency coordination efforts are beginning in the State of New Mexico as well. In 1978, Congress passed the Uranium Mill Tailings Radiation Control Act (UMTRCA), a measure designed to assist in the cleanup of 22 inactive ore-processing sites throughout the southwest. This also included constructing 19 disposal sites for the tailings, which contain a total of 40 million cubic yards of low-level radioactive material. The Environmental Protection Agency estimates that there are 4000 mines with documented uranium production, and another 15,000 locations with uranium occurrences in 14 western states, most found in the Four Corners area and Wyoming.

The *Uranium Mill Tailings Radiation Control Act* is a United States environmental law that amended the Atomic Energy Act of 1954 and gave the Environmental Protection Agency the authority to establish health and environmental standards for the stabilization, restoration, and disposal of uranium mill waste. Title 1 of the Act required the EPA to set environmental protection standards consistent with the Resource Conservation and Recovery Act, including groundwater protection limits; the Department of Energy to implement EPA standards and provide perpetual care for some sites; and the Nuclear Regulatory Commission to review cleanups and license sites to states or the DOE for perpetual care. Title 1 established a uranium mill remedial action program jointly funded by the federal government and the state. Title 1 of the Act also designated 22 inactive uranium mill sites for remediation, resulting in the containment of 40 million cubic yards of low-level radioactive material in UMTRCA Title 1 holding cells.

References

- Lewis, John S. (2015). Asteroid Mining 101: Wealth for the New Space Economy. Deep Space Industries Inc. ISBN 978-0-9905842-0-9. Retrieved 21 May 2015.

- "ASTEROID MINING CORPORATION LIMITED - Overview (free company information from Companies House)". beta.companieshouse.gov.uk. Retrieved 2016-04-01.

- de Selding, Peter B. (2016-06-03). "Luxembourg invests to become the 'Silicon Valley of space resource mining'". SpaceNews. Retrieved 2016-06-04.

- "Nuclear power - the energy balance" (PDF). October 2007. Section D10. Archived from the original (PDF) on November 22, 2008. Retrieved 2016-03-31.

- Lu, Anne (2015-04-21). "Asteroid Mining Could Be The Next Frontier For Resource Mining". International Business Times. Retrieved 23 April 2015.

- Fung, Brian (2015-05-22). "The House just passed a bill about space mining. The future is here.". Washington Post. Retrieved 14 September 2015.

- Rosenbaum, Dr. Helen (November 2011). "Out of Our Depth: Mining the Ocean Floor in Papua New Guinea". Deep Sea Mining Campaign. MiningWatch Canada, CELCoR, Packard Foundation. Retrieved December 2015.

- Palaszweski, Bryan (April 2015). "Atmospheric Mining in the Outer Solar System: Resource Capturing, Exploration, and Exploitation" (PDF). http://gltrs.grc.nasa.gov. Cleveland, Ohio 44135-3191: National Aeronautics and Space Administration John H. Glenn Center at Lewis Field. Retrieved August 13, 2015.

- Komnenic, Ana (7 February 2014). "Paladin Energy suspends production at Malawi uranium mine". Mining.com. Retrieved 17 April 2014.

- Soper, Taylor (January 22, 2013). "Deep Space Industries entering asteroid-mining world, creates competition for Planetary Resources". GeekWire: Dispatches from the Digital Frontier. GeekWire. Retrieved January 22, 2013.

- "Commercial Asteroid Hunters announce plans for new Robotic Exploration Fleet" (Press release). Deep Space Industries. January 22, 2013. Retrieved January 22, 2013.

- Wall, Mike (January 22, 2013). "Asteroid-Mining Project Aims for Deep-Space Colonies". Space.com. TechMediaNetwork. Retrieved January 22, 2013.

- Brendan Drain (23 January 2011). "EVE Evolved: Mining 101 -- Advanced mining". EVE Evolved. Joystiq. Retrieved 12 February 2013.

- MMOGames (20 April 2012). "EVE Online Beginner's Guide - Episode 3 (Choosing A Focus)" (Video). EVE Online Beginner's Guide. YouTube. Retrieved 12 February 2013.

- "Official website for DOE Project Extraction of Uranium from Seawater". Web.ornl.gov. 2012-06-08. Retrieved 2013-07-26.

- "Oak Ridge National Laboratory - ORNL technology moves scientists closer to extracting uranium from seawater". Ornl.gov. 2012-08-21. Retrieved 2013-07-26.

- Mohan, Keerthi (2012-08-13). "New Class of Easily Retrievable Asteroids That Could Be Captured With Rocket Technology Found". International Business Times. Retrieved 2012-08-15.

- Jeffrey Kluger (April 25, 2012). "Can James Cameron — Or Anyone — Really Mine Asteroids?". Time Science. Retrieved 2012-04-25.

Surface Mining and its Techniques

Surface mining is a part of mining. In contrast to underground mining, surface mining is the removal of soil and rock overlying the mineral deposit. The techniques explained in this chapter are open pit mining, mountaintop removal mining, sand mining and dredging.

Surface Mining

Surface mining, including strip mining, open-pit mining and mountaintop removal mining, is a broad category of mining in which soil and rock overlying the mineral deposit (the overburden) are removed. In contrast to underground mining, in which the overlying rock is left in place, and the mineral removed through shafts or tunnels.

Coal strip mine in Wyoming

Surface mining began in the mid-sixteenth century and is practiced throughout the world, although the majority of surface coal mining occurs in North America. It gained popularity throughout the 20th century, and surface mines now produce most of the coal mined in the United States.

In most forms of surface mining, heavy equipment, such as earthmovers, first remove the overburden. Next, huge machines, such as dragline excavators or Bucket wheel excavators, extract the mineral.

Sulfur miner with sulfur obtained from Ijen Volcano, Indonesia (2015)

Types

There are five main forms of surface mining, detailed below.

Strip Mining

The Bagger 288 is a bucket-wheel excavator used in strip mining.

"Strip mining" is the practice of mining a seam of mineral, by first removing a long strip of overlying soil and rock (the overburden). It is most commonly used to mine coal and lignite (brown coal). Strip mining is only practical when the ore body to be excavated is relatively near the surface. This type of mining uses some of the largest machines on earth, including bucket-wheel excavators which can move as much as 12,000 cubic meters of earth per hour.

There are two forms of strip mining. The more common method is "area stripping", which is used on fairly flat terrain, to extract deposits over a large area. As each long strip is excavated, the overburden is placed in the excavation produced by the previous strip.

"Contour stripping" involves removing the overburden above the mineral seam near the outcrop in hilly terrain, where the mineral outcrop usually follows the contour of the land. Contour stripping is often followed by auger mining into the hillside, to remove more of the mineral. This method commonly leaves behind terraces in mountainsides.

Strip mining at Garzweiler, Germany. The lignite being extracted is at left, the removed overburden being placed at right. Note that it is a largely flat mine for a horizontal mineral.

Open-pit Mining

The El Chino mine located near Silver City, New Mexico is an open-pit copper mine.

"Open-pit mining" refers to a method of extracting rock or minerals from the earth through their removal from an open pit or borrow. Although open-pit mining is sometimes mistakenly referred to as "strip mining", the two methods are different.

Mountaintop Removal

"Mountaintop removal mining" (MTR) is a form of coal mining that mines coal seams beneath mountaintops by first removing the mountaintop overlying the coal seam. Explosives are used to break up the rock layers above the seam, which are them removed. Excess mining waste or "overburden" is dumped by large trucks into fills in nearby hollow or valley fills. MTR involves the mass restructuring of earth in order to reach the coal seam as deep as 400 feet (120 m) below the surface. Mountaintop removal replaces the original steep landscape with a much flatter topography. Economic development attempts on reclaimed mine sites include prisons such the Big Sandy Federal Penitentiary in Martin County, Kentucky, small town airports, golf courses such as Twisted Gun in Mingo County, West Virginia and Stonecrest Golf Course in Floyd County, Kentucky, as well as industrial scrubber sludge disposal sites, solid waste landfills, trailer parks, explosive manufacturers, and storage rental lockers.

The technique has been used increasingly in recent years in the Appalachian coal fields of West Virginia, Kentucky, Virginia and Tennessee in the United States. The profound changes in topography and disturbance of pre-existing ecosystems have made mountaintop removal highly controversial.

Advocates of mountaintop removal point out that once the areas are reclaimed as mandated by law, the technique provides premium flat land suitable for many uses in a region where flat land is at a premium. They also maintain that the new growth on reclaimed mountaintop mined areas is better able to support populations of game animals.

Critics contend that mountaintop removal is a disastrous practice that benefits a small number of corporations at the expense of local communities and the environment. A U.S. Environmental Protection Agency (EPA) environmental impact statement finds that streams near valley fills sometimes may contain higher levels of minerals in the water and decreased aquatic biodiversity.

The statement also estimates that 724 miles (1,165 km) of Appalachian streams were buried by valley fills from 1985 to 2001.

Blasting at a mountaintop removal mine expels dust and fly-rock into the air, which can then disturb or settle onto private property nearby. This dust may contain sulfur compounds, which some claim corrode structures and tombstones and is a health hazard.

Although MTR sites are required to be reclaimed after mining is complete, reclamation has traditionally focused on stabilizing rock and controlling erosion, but not always on reforesting the area. Quick-growing, non-native grasses, planted to quickly provide vegetation on a site, compete with tree seedlings, and trees have difficulty establishing root systems in compacted backfill. Consequently, biodiversity suffers in a region of the United States with numerous endemic species. Erosion also increases, which can intensify flooding. In the Eastern United States, the Appalachian Regional Reforestation Initiative works to promote the use of trees in mining reclamation.

Dredging

"Dredging" is a method for placer mining below the water table. It is mostly associated with gold mining. Small dredges often use suction to bring the mined material up from the bottom of a water body. Historically, large-scale dredging often used a floating dredge, a barge-like vessel which scooped material up on a conveyor belt in front, removed the desirable component on board, and returned the unwanted material via another conveyor belt in back. In gravel-filled river valleys with shallow water tables, a floating dredge could work its way through the loose sediment in a pond of its own making.

Highwall Mining

Highwall mining is another form of surface mining that evolved from auger mining. In Highwall mining, the coal seam is penetrated by a continuous miner propelled by a hydraulic Pushbeam Transfer Mechanism (PTM). A typical cycle includes sumping (launch-pushing forward) and shearing (raising and lowering the cutterhead boom to cut the entire height of the coal seam). As the coal recovery cycle continues, the cutterhead is progressively launched into the coal seam for 19.72 feet (6.01 m). Then, the Pushbeam Transfer Mechanism (PTM) automatically inserts a 19.72-foot (6.01 m) long rectangular Pushbeam (Screw-Conveyor Segment) into the center section of the machine between the Powerhead and the cutterhead. The Pushbeam system can penetrate nearly 1,000 feet (300 m) into the coal seam. One patented Highwall mining systems use augers enclosed inside the Pushbeam that prevent the mined coal from being contaminated by rock debris during the conveyance process. Using a video imaging and/or a gamma ray sensor and/or other Geo-Radar systems like a coal-rock interface detection sensor (CID), the operator can see ahead projection of the seam-rock interface and guide the continuous miner's progress. Highwall mining can produce thousands of tons of coal in contour-strip operations with narrow benches, previously mined areas, trench mine applications and steep-dip seams with controlled water-inflow pump system and/or a gas (inert) venting system.

Recovery is much better than Augering, but the mapping of areas that have been developed by a Highwall miner are not mapped as rigorously as deep mined areas. Very little soil is displaced in contrast with mountain top removal; however a large amount of capital is required to operate and own a Highwall miner. But then this Highwall mining system is the innovative roadmap future potential and stay or being better competitive in the area of environmental friendly non mountain-top (overburden) removal operated by only 4 crew members.

Mapping of the outcrop as well as core hole data and samples taken during the bench making process are taken into account to best project the panels that the Highwall miner will cut. Obstacles that could be potentially damaged by subsidence and the natural contour of the Highwall are taken into account, and a surveyor points the Highwall miner in a line (Theoretical Survey Plot-Line) mostly perpendicular to the Highwall. Parallel lines represent the drive cut into the mountain (up to 1,000 feet (300 m) deep), without heading or corrective steering actuation on a navigation Azimuth during mining results in missing a portion of the coal seam and is a potential danger of cutting in pillars from previous mined drives due to horizontal drift (Roll) of the Pushbeam-Cuttermodule string. Recently Highwall miners have penetrated more than 1050 feet into the coal seam, and today's models are capable of going farther, with the support of gyro navigation and not limited anymore by the amount of cable stored on the machine. The maximum depth would be determined by the stress of further penetration and associated specific-power draw, ("Torsion and Tension" in Screw-Transporters String) but today's optimized Screw-Transporters Conveying Embodiments (called: Pushbeams) with Visual Product Development and Flow Simulation Behaviour software "Discrete Element Modeling" (DEM) shows smart-drive extended penetrations are possible, evenso under steep inclined angles from horizontal to more than 30 degree downhole. In case of significant steep mining the new mining method phrase should be "Directional Mining", dry or wet, Dewatering is developed or Cutting & Dredging through Screw-Transporters are proactive in developing roadmap of the leading global Highwall mining company.

Environmental and Health Issues

The impact of surface mining on the topography, vegetation, and water resources has made it highly controversial.

Surface mining is subject to state and federal reclamation requirements, but adequacy of the requirements is a constant source of contention. Unless reclaimed, surface mining can leave behind large areas of infertile waste rock, as 70% of material excavated is waste.

In the United States, the Surface Mining Control and Reclamation Act of 1977 mandates reclamation of surface coal mines. Reclamation for non-coal mines is regulated by state and local laws, which may vary widely.

Human Health

The United Mine Workers of America has spoken against the use of human sewage sludge to reclaim surface mining sites in Appalachia. The UMWA launched its campaign against the use of sludge on mine sites in 1999 after eight UMWA workers became ill from exposure to Class B sludge spread near their workplace.

Environmental Impact

According to a 2010 report in the journal *Science*, mountaintop mining has caused numerous environmental problems which mitigation practices have not successfully addressed. For example, valley fills frequently bury headwater streams causing permanent loss of ecosystems. In addition, the destruction of large tracts of deciduous forests has threatened several endangered species and led to a loss of biodiversity.

Open-pit Mining

Rock blasting at the large open-pit Twin Creeks gold mine in Nevada, United States. Note the size of the excavators for scale (foreground, left), and that the bottom of the mine is not visible.

Open-pit, open-cast or open cut mining is a surface mining technique of extracting rock or minerals from the earth by their removal from an open pit or borrow.

This form of mining differs from extractive methods that require tunneling into the earth, such as long wall mining. Open-pit mines are used when deposits of commercially useful minerals or rocks are found near the surface; that is, where the overburden (surface material covering the valuable deposit) is relatively thin or the material of interest is structurally unsuitable for tunneling (as would be the case for sand, cinder, and gravel). For minerals that occur deep below the surface—where the overburden is thick or the mineral occurs as veins in hard rock—underground mining methods extract the valued material.

An open-pit copper mine in Chuquicamata.

Open-pit mines that produce building materials and dimension stone are commonly referred to as "quarries."

Open-pit mines are typically enlarged until either the mineral resource is exhausted, or an increasing ratio of overburden to ore makes further mining uneconomic. When this occurs, the exhausted mines are sometimes converted to landfills for disposal of solid wastes. However, some form of water control is usually required to keep the mine pit from becoming a lake, if the mine is situated in a climate of considerable precipitation or if any layers of the pit forming the mine border productive aquifers.

Extraction

Note the angled and stepped sides of the Sunrise Dam Gold Mine, Australia.

The Klunst rock quarry in Saxony, Germany, has few benches.

Open-cast mines are dug on benches, which describe vertical levels of the hole. These benches are usually on four to sixty meter intervals, depending on the size of the machinery that is being used. Many quarries do not use benches, as they are usually shallow.

Most walls of the pit are generally dug on an angle less than vertical, to prevent and minimize damage and danger from rock falls. This depends on how weathered the rocks are (eroded rocks), and the type of rock, and also how many structural weaknesses occur within the rocks, such as a faults, shears, joints or foliations.

The walls are stepped. The inclined section of the wall is known as the batter, and the flat part of the step is known as the bench or berm. The steps in the walls help prevent rock falls continuing down the entire face of the wall. In some instances additional ground support is required and rock bolts, cable bolts and shotcrete are used. De-watering bores may be used to relieve water pressure by drilling horizontally into the wall, which is often enough to cause failures in the wall by itself.

A haul road is usually situated at the side of the pit, forming a ramp up which trucks can drive, carrying ore and waste rock.

Waste rock is piled up at the surface, near the edge of the open pit. This is known as the waste dump. The waste dump is also tiered and stepped, to minimize degradation.

Ore which has been processed is known as tailings, and is generally a slurry. This is pumped to a tailings dam or settling pond, where the water evaporates. Tailings dams can often be toxic due to the presence of unextracted sulfide minerals, some forms of toxic minerals in the gangue, and often cyanide which is used to treat gold ore via the cyanide leach process. This toxicity can harm the surrounding environment.

Rehabilitation

Opencut coal mine loadout station and reclaimed land at the North Antelope Rochelle coal mine in Wyoming, United States.

After mining finishes, the mine area may undergo land rehabilitation. Waste dumps are contoured to flatten them out, to further stabilise them. If the ore contains sulfides it is usually covered with a layer of clay to prevent access of rain and oxygen from the air, which can oxidise the sulfides to produce sulfuric acid, a phenomenon known as acid mine drainage. This is then generally covered with soil, and vegetation is planted to help consolidate the material. Eventually this layer will erode, but it is generally hoped that the rate of leaching or acid will be slowed by the cover such that the environment can handle the load of acid and associated heavy metals. There are no long term studies on the success of these covers due to the relatively short time in which large scale open pit mining has existed. It may take hundreds to thousands of years for some waste dumps to become "acid neutral" and stop leaching to the environment. The dumps are usually fenced off to prevent livestock denuding them of vegetation. The open pit is then surrounded with a fence, to prevent access, and it generally eventually fills up with ground water. In arid areas it may not fill due to deep groundwater levels. Instead of returning the land to its former natural state, it may also be reused, converting it into recreational parks or even residential/mixed communities.

An open-pit sulfur mine at Tarnobrzeg, Poland undergoing land rehabilitation

Typical Open Cut Grades

Gold is generally extracted in open-pit mines at 1 to 2 ppm (parts per million) but in certain cases, 0.75 ppm gold is economical. This was achieved by bulk heap leaching at the Peak Hill mine in western New South Wales, near Dubbo, Australia.

Nickel, generally as laterite, is extracted via open-pit down to 0.2%. Copper is extracted at grades as low as 0.15% to 0.2%, generally in massive open-pit mines in Chile, where the size of the resources and favorable metallurgy allows economies of scale.

Mountaintop Removal Mining

Mountaintop removal site

Mountaintop removal site in Pike County, Kentucky

Mountaintop removal mining (MTR), also known as mountaintop mining (MTM), is a form of surface mining that involves the mining of the summit or summit ridge of a mountain. Coal seams are extracted from a mountain by removing the land, or overburden, above the seams. This method of coal mining is conducted in the Appalachian Mountains in the eastern United States. Explosives are used to remove up to 400 vertical feet (120 m) of mountain to expose underlying coal seams. Excess rock and soil is dumped into nearby valleys, in what are called "holler fills" or "valley fills." Less expensive to execute and requiring fewer employees, mountaintop removal mining began in Appalachia in the 1970s as an extension of conventional strip mining techniques. It is primarily occurring in Kentucky, West Virginia, Virginia, and Tennessee.

The practice of mountaintop removal mining has been controversial. The coal industry cites economic benefits and asserts that mountaintop removal is safer than underground mining. Published scientific studies have found that mountaintop mining has serious environmental impacts that mitigation practices cannot successfully address. A high potential for human health impacts has also been reported.

Overview

Mountaintop removal mining (MTR), also known as mountaintop mining (MTM), is a form of surface mining that involves the topographical alteration and/or removal of a summit, hill, or ridge to access buried coal seams.

The MTR process involves the removal of coal seams by first fully removing the overburden laying atop them, exposing the seams from above. This method differs from more traditional underground mining, where typically a narrow shaft is dug which allows miners to collect seams using various underground methods, while leaving the vast majority of the overburden undisturbed. The overburden from MTR is either placed back on the ridge, attempting to reflect the approximate original contour of the mountain, and/or it is moved into neighboring valleys.

Excess rock and soil containing mining byproducts are disposed into nearby valleys, in what are called "holler fills" or "valley fills."

MTR in the United States is most often associated with the extraction of coal in the Appalachian Mountains, where the United States Environmental Protection Agency (EPA) estimates that 2,200 square miles (5,700 km²) of Appalachian forests will be cleared for MTR sites by the year 2012. Sites range from Ohio to Virginia. It occurs most commonly in West Virginia and Eastern Kentucky, the top two coal-producing states in Appalachia, with each state using approximately 1,000 tonnes of explosives per day for surface mining. At current rates, MTR in the U.S. will mine over 1.4 million acres (5,700 km²) by 2010, an amount of land area that exceeds that of the state of Delaware.

Mountaintop removal has been practiced since the 1960s. Increased demand for coal in the United States, sparked by the 1973 and 1979 petroleum crises, created incentives for a more economical form of coal mining than the traditional underground mining methods involving hundreds of workers, triggering the first widespread use of MTR. Its prevalence expanded further in the 1990s to retrieve relatively low-sulfur coal, a cleaner-burning form, which became desirable as a result of amendments to the U.S. Clean Air Act that tightened emissions limits on high-sulfur coal processing.

Process

US EPA diagram of mountaintop mining:

"Step 1. Layers of rock and dirt above the coal (called overburden) are removed."
"Step 2. The upper seams of coal are removed with spoils placed in an adjacent valley." "Step 3. Draglines excavate lower layers of coal with spoils placed in spoil piles."
"Step 4. Regrading begins as coal excavation continues."
"Step 5. Once coal removal is completed, final regrading takes place and the area is revegetated."

Land is deforested prior to mining operations and the resultant lumber is either sold or burned. According to the Surface Mining Control and Reclamation Act of 1977 (SMCRA), the topsoil is supposed to be removed and set aside for later reclamation. However, coal companies are often granted waivers and instead reclaim the mountain with "topsoil substitute." The waivers are granted if adequate amounts of topsoil are not naturally present on the rocky ridge top. Once the area is cleared, miners use explosives to blast away the overburden, the rock and subsoil, to expose coal seams beneath. The overburden is then moved by various mechanical means to areas of the ridge previously mined. These areas are the most economical area of storage as they are located close to the active pit of exposed coal. If the ridge topography is too steep to adequately handle the amount of spoil produced then additional storage is used in a nearby valley or hollow, creating what is known as a *valley fill* or *hollow fill*. Any streams in a valley are buried by the overburden.

A front-end loader or excavator then removes the coal, where it is transported to a processing plant. Once coal removal is completed, the mining operators back stack overburden from the next area to be mined into the now empty pit. After backstacking and grading of overburden has been completed, topsoil (or a topsoil substitute) is layered over the overburden layer. Next, grass seed is spread in a mixture of seed, fertilizer, and mulch made from recycled newspaper. Depending on surface land owner wishes the land will then be further reclaimed by adding trees if the pre-approved post-mining land use is forest land or wildlife habitat. If the land owner has requested other post-mining land uses the land can be reclaimed to be used as pasture land, economic development or other uses specified in SMCRA.

Because coal usually exists in multiple geologically stratified seams, miners can often repeat the blasting process to mine over a dozen seams on a single mountain, increasing the mine depth each time. This can result in a vertical descent of hundreds of extra feet into the earth.

Economics

Almost half of the electricity generated in the United States is produced by coal-fired power plants. MTR accounted for less than 5% of U.S. coal production as of 2001. In some regions, however, the percentage is higher, for example MTR provided 30% of the coal mined in West Virginia in 2006.

Historically in the U.S. the prevalent method of coal acquisition was underground mining which is very labor-intensive. In MTR, through the use of explosives and large machinery, more than two and a half times as much coal can be extracted per worker per hour than in traditional underground mines, thus greatly reducing the need for workers. In Kentucky, for example, the number of workers has declined over 60% from 1979 to 2006 (from 47,190 to 17,959 workers). The industry overall lost approximately 10,000 jobs from 1990 to 1997, as MTR and other more mechanized underground mining methods became more widely used. The coal industry asserts that surface mining techniques, such as mountaintop removal, are safer for miners than sending miners underground.

Proponents argue that in certain geologic areas, MTR and similar forms of surface mining allow the only access to thin seams of coal that traditional underground mining would not be able to mine. MTR is sometimes the most cost-effective method of extracting coal.

Several studies of the impact of restrictions to mountaintop removal were authored in 2000 through 2005. Studies by Mark L. Burton, Michael J. Hicks and Cal Kent identified significant state level tax losses attributable to lower levels of mining (notably the studies did not examine potential environmental costs, which the authors acknowledge may outweigh commercial benefits). Mountaintop removal sites are normally restored after the mining operation is complete, but "reclaimed soils characteristically have higher bulk density, lower organic content, low water-infiltration rates, and low nutrient content.

Legislation in the United States

In the United States, MTR is allowed by section 515(c)(1) of the Surface Mining Control and Reclamation Act of 1977. Although most coal mining sites must be reclaimed to the land's pre-mining contour and use, regulatory agencies can issue waivers to allow MTR. In such cases, SMCRA dictates that reclamation must create "a level plateau or a gently rolling contour with no highwalls remaining."

Permits must be obtained to deposit valley fill into streams. On four occasions, federal courts have ruled that the US Army Corps of Engineers violated the Clean Water Act by issuing such permits. Massey Energy Company is currently appealing a 2007 ruling, but has been allowed to continue mining in the meantime because "most of the substantial harm has already occurred," according to the judge.

The Bush administration appealed one of these rulings in 2001 because the Act had not explicitly defined "fill material" that could legally be placed in a waterway. The EPA and Army Corps of Engineers changed a rule to include mining debris in the definition of fill material, and the ruling was overturned.

On December 2, 2008, the Bush Administration made a rule change to remove the Stream Buffer Zone protection provision from SMCRA allowing coal companies to place mining waste rock and dirt directly into headwater waterways.

A federal judge has also ruled that using settling ponds to remove mining waste from streams violates the Clean Water Act. He also declared that the Army Corps of Engineers has no authority to issue permits allowing discharge of pollutants into such in-stream settling ponds, which are often built just below valley fills.

On January 15, 2008, the environmental advocacy group Center for Biological Diversity petitioned the United States Fish and Wildlife Service (FWS) to end a policy that waives detailed federal Endangered Species Act reviews for new mining permits. The current policy states that MTR can never damage endangered species or their habitat as long as mining operators comply with federal surface mining law, despite the complexities of species and ecosystems. Since 1996, this policy has exempted many strip mines from being subject to permit-specific reviews of impact on individual endangered species. Because of the 1996 Biological Opinion by FWS making case-by-case formal reviews unnecessary, the Interior's Office of Surface Mining and state regulators require mining companies to hire a government-approved contractor to conduct their own surveys for any potential endangered species. The surveys require approval from state and federal biologists, who provide informal guidance on how to minimize mines' potential effects to species. While the agencies have the option to ask for formal endangered species consultations during that process, they do so very rarely.

On May 25, 2008, North Carolina State Representative Pricey Harrison introduced a bill to ban the use of mountaintop removal coal from coal-fired power plants within North Carolina. This proposed legislation would have been the only legislation of its kind in the United States; however, the bill was defeated.

Environmental and Health Impacts

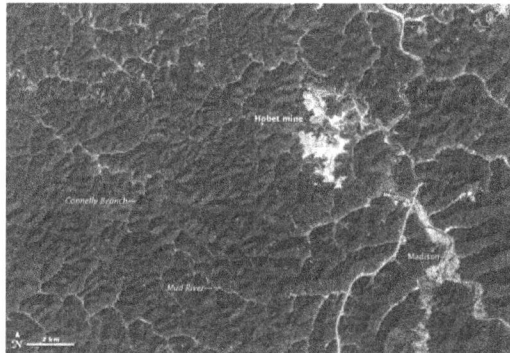

The Hobet mine in West Virginia taken by NASA LANDSAT in 1984

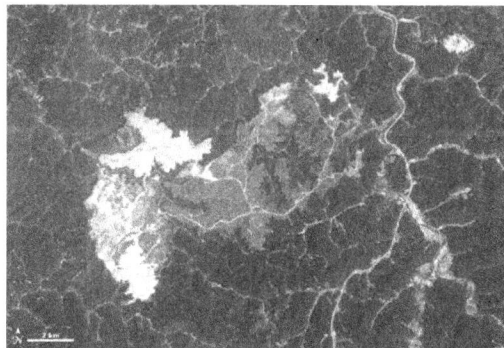

The Hobet mine in West Virginia taken by NASA LANDSAT in 2009

Critics contend that MTR is a destructive and unsustainable practice that benefits a small number of corporations at the expense of local communities and the environment. Though the main issue has been over the physical alteration of the landscape, opponents to the practice have also criticized MTR for the damage done to the environment by massive transport trucks, and the environmental damage done by the burning of coal for power. Blasting at MTR sites also expels dust and fly-rock into the air, which can disturb or settle onto private property nearby. This dust may contain sulfur compounds, which corrodes structures and is a health hazard.

A January 2010 report in the journal *Science* reviews current peer-reviewed studies and water quality data and explores the consequences of mountaintop mining. It concludes that mountaintop mining has serious environmental impacts that mitigation practices cannot successfully address. For example, the extensive tracts of deciduous forests destroyed by mountaintop mining support several endangered species and some of the highest biodiversity in North America. There is a particular problem with burial of headwater streams by valley fills which causes permanent loss of ecosystems that play critical roles in ecological processes. In addition, increases in metal ions, pH, electrical conductivity, total dissolved solids due to elevated concentrations of sulfate are closely linked to the extent of mining in West Virginia watersheds. Declines in stream biodiversity have been linked to the level of mining disturbance in West Virginia watersheds.

Published studies also show a high potential for human health impacts. These may result from contact with streams or exposure to airborne toxins and dust. Adult hospitalization for chronic pulmonary disorders and hypertension are elevated as a result of county-level coal production. Rates of mortality, lung cancer, as well as chronic heart, lung and kidney disease are also increased. A 2011 study found that counties in and near mountaintop mining areas had higher rates of birth defects for five out of six types of birth defects, including circulatory/respiratory, musculoskeletal, central nervous system, gastrointestinal, and urogenital defects. These defect rates were more pronounced in the most recent period studied, suggesting the health effects of mountaintop mining-related air and water contamination may be cumulative. Another 2011 study found "the odds for reporting cancer were twice as high in the mountaintop mining environment compared to the non mining environment in ways not explained by age, sex, smoking, occupational exposure, or family cancer history."

A United States Environmental Protection Agency (EPA) environmental impact statement finds that streams near some valley fills from mountaintop removal contain higher levels of minerals in the water and decreased aquatic biodiversity. Mine-affected streams also have high selenium concentrations, which can bioaccumulate and produce toxic effects (e.g., reproductive failure, physical deformity, mortality), and these effects have been documented in reservoirs below streams (Lemly 2008). The statement also estimates that 724 miles (1,165 km) of Appalachian streams were buried by valley fills between 1985 and 2001. On September 28, 2010, the U.S. Environmental Protection Agency's (EPA) independent Science Advisory Board (SAB) released their first draft review of EPA's research into the water quality impacts of valley fills associated with mountaintop mining, agreeing with EPA's conclusion that valley fills are associated with increased levels of conductivity threatening aquatic life in surface waters.

Although U.S. mountaintop removal sites by law must be reclaimed after mining is complete, reclamation has traditionally focused on stabilizing rock formations and controlling for erosion, and not on the reforestation of the affected area. Fast-growing, non-native flora such as *Lespedeza cuneata*, planted to quickly provide vegetation on a site, compete with tree seedlings, and trees have difficulty establishing root systems in compacted backfill. Consequently, biodiversity suffers in a region of the United States with numerous endemic species. In addition, reintroduced elk (*Cervus canadensis*) on mountaintop removal sites in Kentucky are eating tree seedlings.

Advocates of MTR claim that once the areas are reclaimed as mandated by law, the area can provide flat land suitable for many uses in a region where flat land is at a premium. They also maintain that the new growth on reclaimed mountaintop mined areas is better suited to support populations of game animals. While some of the land is able to be turned into grassland which game animals can live in, the amount of grassland is minimal. The land does not retake the form it had before the MTR. As stated in the book *Bringing Down the Mountains*: "Some of the main problems associated with MTR include soil depletion, sedimentation, low success rate of tree regrowth, lack of successful revegetation, displacement of native wildlife, and burial of streams." The ecological benefits after MTR are far below the level of the original land.

Art, Entertainment, and Media

Documentaries

- Catherine Pancake released the first comprehensive, feature-length documentary on mountaintop removal, *Black Diamonds: Mountaintop Removal and the Search for Coalfield Justice* (2006), a selection in the Documentary Fortnight at the Museum of Modern Art. The film features Julia Bonds, who won the 2003 Goldman Environmental Prize.

- The documentary, *Mountain Top Removal* (2007), focuses on Mountain Justice Summer activists, coal field residents, and coal industry officials. On April 18, 2008, the film received the Reel Current award selected and presented by Al Gore at the Nashville Film Festival.

- The feature documentary, *Burning the Future: Coal in America* (2008), was awarded the International Documentary Association's 2008 Pare Lorentz award for Best Documentary.

- *The Last Mountain* (2011), directed by Bill Haney, details the effects on the land and people living near mountaintop removal and coal burning sites. Maria Gunnoe, the 2009 Goldman Environmental Prize winner, Robert F. Kennedy, Jr., and others present the devastation, confront the politicians and corporate interests, and offer wind power as one solution for Coal River Mountain, West Virginia.

- The autoethnographic documentary film *Goodbye Gauley Mountain: An Ecosexual Love Story* (2013), by Beth Stephens with Annie Sprinkle, raises awareness on the issue of mountain top removal in West Virginia by bringing together environmental activism, performance art, and queer activism against the issue. Stephens says: "My hope for this film, is that in addition to it being a compelling story, it will inspire and raise awareness in groups of people not normally associated with the environmental movement, and especially in LGBTQ communities. There are relatively few films about environmental issues that feature out queers."

Non-fiction Books

- In April 2005, a group of Kentucky writers traveled together to see the devastation from mountaintop removal mining, and Wind Publishing produced the resulting collection of poems, essays and photographs, co-edited by Kristin Johannesen, Bobbie Ann Mason, and Mary Ann Taylor-Hall in *Missing Mountains: We went to the mountaintop, but it wasn't there.*

- Dr. Shirley Stewart Burns, a West Virginia coalfield native, wrote the first academic work on mountaintop removal, titled *Bringing Down The Mountains* (2007), which is loosely based on her internationally award-winning 2005 Ph.D. dissertation of the same name.

- Dr. Burns was also a co-editor, with Kentucky author Silas House and filmmaker Mari-Lynn Evans, of *Coal Country* (2009), a companion book for the nationally recognized feature-length film of the same name.

- *House, Silas & Howard, Jason (2009). Something's Rising: Appalachians Fighting Mountaintop Removal. Lexington, KY: The University Press of Kentucky. ISBN 978-0-8131-2546-6.*

- *Howard, Jason (Editor) (2009). We All Live Downstream: Writings about Mountaintop Removal. Louisville, KY: Motes Books. ISBN 978-1-934894-07-1.*

- Dr. Rebecca Scott, another native West Virginian, examined the sociological relationship of identity and natural resource extraction in central Appalachia in her book, *Removing Mountains* (2010).

- *Hedges, Chris & Sacco, Joe (2012). Days of Destruction, Days of Revolt. Nation Books. ISBN 1568586434.*

- Cultural historian Jeff Biggers published *The United States of Appalachia* (), which examined the cultural and human costs of mountaintop removal.

Additionally, many personal interest stories of coalfield residents have been written, including:

- *Lost Mountain* (2007) by Erik Reese

- *Moving Mountains: How One Woman and Her Community Won Justice From Big Coal* (2007) by Penny Loeb

Fiction Books

- Ann Pancake's *Strange As This Weather Has Been* (2007), is a novel about the subject.

- Mountaintop removal is a major plot element of Jonathan Franzen's best-selling novel *Freedom* (2010), wherein a major character helps to secure land for surface mining with the promise that it will be restored and turned into a nature reserve.

- *Same Sun Here* by Silas House and Neela Vaswani is a novel for middle grade readers that deals with issues of mountaintop removal and is set over the course of one school year 2008-2009.

- In John Grisham's novel *Gray Mountain* (2014), Samantha Kofer moves from a large Wall Street law firm to a small Appalachian town where she confronts the world of coal mining.

Music

- Caroline Herring's song *Black Mountain Lullaby* (on the album *Camilla*, 2012) is based on the story of Jeremy Davidson, age 3, who was killed by a mountaintop mining accident in 2004. She was inspired to write the song after reading an editorial about mountaintop removal written by Silas House that appeared in the New York Times on 19 February 2011.

- Lissie's album 'Back To Forever' contains a moving protest song on the topic called simply 'Mountaintop Removal'.

- Liam Wilson of The Dillinger Escape Plan wore a homemade shirt saying "stop mtm/vf" during the bands performance on Late Night with Conan O'Brien.

Sand Mining

Sand mining is a practice that is used to extract sand, mainly through an open pit. However, sand is also mined from beaches, inland dunes and dredged from ocean beds and river beds. It is often used in manufacturing as an abrasive, for example, and it is used to make concrete. It is also used in cold regions to put on the roads by municipal plow trucks to help icy and snowy driving conditions, usually mixed with salt or another mixture to lower the freezing temperature of the road surface (have the precipitations freeze at a lower temperature). Sand dredged from the mouths of rivers can also be used to replace eroded coastline.

Another reason for sand mining is the extraction of minerals such as rutile, ilmenite and zircon, which contain the industrially useful elements titanium and zirconium. These minerals typically occur combined with ordinary sand, which is dug up, the valuable minerals being separated in water by virtue of their different densities, and the remaining ordinary sand re-deposited.

Sand mining is a direct cause of erosion, and also impacts the local wildlife. For example, sea turtles depend on sandy beaches for their nesting, and sand mining has led to the near extinction of gharials (a species of crocodiles) in India. Disturbance of underwater and coastal sand causes turbidity in the water, which is harmful for such organisms as corals that need sunlight. It also destroys fisheries, causing problems for people who rely on fishing for their livelihoods.

Removal of physical coastal barriers such as dunes leads to flooding of beachside communities, and the destruction of picturesque beaches causes tourism to dissipate. Sand mining is regulated by law in many places, but is still often done illegally.

By Country

Sand mine in the Czech Republic.

Australia

In the 1940 mining operations began on the Kurnell Peninsula (Captain Cook's landing place in Australia) to supply the expanding Sydney building market. It continued until 1990 with an estimate of over 70 million tonnes of sand having been removed. The sand has been valued for many decades by the building industry, mainly because of its high crushed shell content and lack of organic matter, it has provided a cheap source of sand for most of Sydney since sand mining operations began. The site has now been reduced to a few remnant dunes and deep water-filled pits which are now being filled with demolition waste from Sydney's building sites. Removal of the sand has significantly weakened the peninsula's capacity to resist storms. Ocean waves pounding against the reduced Kurnell dune system have threatened to break through to Botany Bay, especially during the storms of May and June back in 1974 and of August 1998. Sand Mining also takes place in the Stockton sand dunes north of Newcastle and in the Broken Hill region in the far west of the state.

A large and long running sand mine in Queensland, Australia (on North Stradbroke Island) provides a case study in the (disastrous) environmental consequences on a fragile sandy-soil based ecosystem, justified by the provision of low wage casual labor on an island with few other work options. The Labor state government pledged to end sandmining by 2025, but this decision was overturned by the LNP government which succeeded it. This decision has been subject to allegation of corrupt conduct.

Sand mining contributes to the construction of buildings and development. However, the negative effects of sand mining include the permanent loss of sand in areas, as well as major habitat destruction.

India

Sand mining is a practice that is becoming an environmental issue in India. Environmentalists have raised public awareness of illegal sand mining in the state of Maharashtra and Goa of India. Conservation and environmental NGO Awaaz Foundation filed a public interest litigation in the Bombay High Court seeking a ban on mining activities along the Konkan coast. Awaaz Foundation, in partnership with the Bombay Natural History Society also presented the issue of sand mining as a major international threat to coastal biodiversity at the Conference of Parties 11, Convention on Biological Diversity, Hyderabad in October 2012. D. K. Ravi, an Indian Administrative Service officer of the Karnataka state, who was well known for his tough crackdown on the rampant illegal sand mining in the Kolar district, was found dead at his residence in Bengaluru, on March 16, 2015. It is widely alleged that the death is not due to suicide but the handiwork of the mafia involved in land grabbing and sand mining.

New Zealand

Sand mining occurs in the Kaipara Harbour, off the coast at Pakiri and offshore from Little Barrier Island. A sand mine had operated at Whiritoa on the east coast of the North Island for 50 years extracting 180,000m³ of sand. Coastal sand mines currently operate at Maioro and Taharoa to recover iron sand. When an application was lodged in 2005 to mine iron sands on the seabed of the coast of Raglan local residents organised in opposition to the scheme.

The application for the mining was turned down by Crown Minerals due to a lack of technical detail.

Sierra Leone

Recently, activists and local villagers have protested against sand mining on Sierra Leone's Western Area Peninsular. The activity is contributing to Sierra Leone's coastal erosion, which is proceeding at up to 6 meters a year.

United States

The current size of the sand mining market in the United States is slightly over a billion dollars per year. The industry has been growing by nearly 10% annually since 2005 because of its use in hydrocarbon extraction. The majority of the market size for mining is held by Texas and Illinois.

Wisconsin, Minnesota, Illinois, Indiana and Iowa

Silica sand mining business has more than doubled since 2009 because of the need for this particular type of sand, which is used in a process known as hydraulic fracturing. Wisconsin is one of the five states that produce nearly 2/3 of the nation's silica. As of 2009, Wisconsin, along with other northern states, is facing an industrial mining boom, being dubbed the "sand rush" because of the new demand from large oil companies for silica sand. According to Minnesota Public Radio, "One of the industry's major players, U.S. Silica, says its sand sales tied to hydraulic fracturing nearly doubled to $70 million from 2009 to 2010 and brought in nearly $70 million in *just the first nine months* of 2011." According to the Wisconsin Department of Natural Resources (WDNR), there are currently 34 active mines and 25 mines in development in Wisconsin. In 2012, the WDNR released a final report on the silica sand mining in Wisconsin titled *Silica Sand Mining in Wisconsin*. The recent boom in silica sand mining has caused concern from residents in Wisconsin that include quality of life issues and the threat of silicosis. However, these are issues that the state has no authority to regulate. According to the WDNR (2012) these issues include noise, lights, hours of operation, damage and excessive wear to roads from trucking traffic, public safety concerns from the volume of truck traffic, possible damage and annoyance resulting from blasting, and concerns regarding aesthetics and land use changes.

As of 2013, industrial frac sand mining has become a cause for activism, especially in the Driftless Area of southeast Minnesota, northeast Iowa and southwest Wisconsin.

Dredging

Dredging is an excavation activity usually carried out underwater, in shallow seas or freshwater areas with the purpose of gathering up bottom sediments and disposing of them at a different location. This technique is often used to keep waterways navigable. It is also used as a way to replenish sand on some public beaches, where sand has been lost because of coastal erosion. Dredging is also used as a technique for fishing for certain species of edible clams and crabs.

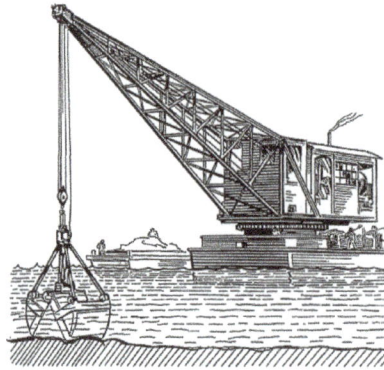

A grab dredge

Uses

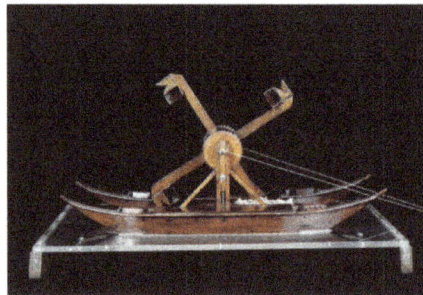

Reconstruction of the mud-drag by Leonardo da Vinci (*Manuscript E, folio 75 v.*) exposed at the Museo nazionale della scienza e della tecnologia "Leonardo da Vinci", Milan.

Reconstruction of the mud-drag by Leonardo da Vinci (*Manuscript E, folio 75 v.*) exposed at the Museo nazionale della scienza e della tecnologia "Leonardo da Vinci", Milan.

- Capital: dredging carried out to create a new harbor, berth or waterway, or to deepen existing facilities in order to allow larger ships access. Because capital works usually involve hard material or high-volume works, the work is usually done using a cutter suction dredge or large trailing suction hopper dredge; but for rock works, drilling and blasting along with mechanical excavation may be used.

- Preparatory: work and excavation for future bridges, piers or docks/wharves, often connected with foundation work.

- Maintenance: dredging to deepen or maintain navigable waterways or channels which are threatened to become silted with the passage of time, due to sedimented sand and mud, possibly making them too shallow for navigation. This is often carried out with a trailing suction hopper dredge. Most dredging is for this purpose, and it may also be done to maintain the holding capacity of reservoirs or lakes.

- Land reclamation: dredging to mine sand, clay or rock from the seabed and using it to construct new land elsewhere. This is typically performed by a cutter-suction dredge or trailing suction hopper dredge. The material may also be used for flood or erosion control.

- Beach nourishment: mining sand offshore and placing on a beach to replace sand eroded by storms or wave action. This is done to enhance the recreational and protective function of the beaches, which can be eroded by human activity or by storms. This is typically performed by a cutter-suction dredge or trailing suction hopper dredge.

- Harvesting materials: dredging sediment for elements like gold, diamonds or other valuable trace substances.

- Seabed mining: a possible future use, recovering natural metal ore nodules from the sea's abyssal plains.

- Construction materials: dredging sand and gravels from offshore licensed areas for use in construction industry, principally for use in concrete. Very specialist industry focused in NW Europe using specialized trailing suction hopper dredgers self discharging dry cargo ashore.

- Anti-eutrophication: Dredging is an expensive option for the remediation of eutrophied (or de-oxygenated) water bodies. However, as artificially elevated phosphorus levels in the sediment aggravate the eutrophication process, controlled sediment removal is occasionally the only option for the reclamation of still waters.

- Contaminant remediation: to reclaim areas affected by chemical spills, storm water surges (with urban runoff), and other soil contaminations. Disposal becomes a proportionally large factor in these operations.

- Removing trash and debris: often done in combination with maintenance dredging, this process removes non-natural matter from the bottoms of rivers and canals and harbors.

- Flood prevention: this can help to increase channel depth and therefore increase a channel's capacity for carrying water.

- Peat extraction: in former times, so-called *dredging poles* or *dredge hauls* were used on the back of small boats to manually dredge the beds of peat-moor waterways before extracting the peat for use as a fuel. This tradition has now become more or less obsolete and the tools used to do this have also changed significantly.

- Oyster dredging or harvesting: in Louisiana and other states with salt water estuaries that can sustain bottom oyster beds. A heavy metal rectangular scoop device is towed astern of a moving boat with a chain bridle attached to a cable and winch which scoops up oysters as it drags along the bottom. The device is periodically hauled aboard and the oysters in it are sorted and bagged for shipment to an oyster processing facility.

- As a hobby: hobbyists examine their dredged matter to pick out items of potential value, similar to the hobby of metal detecting, or the hobby form of dumpster diving, on land.

Relevance

Without the many and almost non-stop dredging operations worldwide, much of the world's commerce would be impaired, often within a few months, since much of world's goods travel by ship, and need to access harbours or seas via channels. Recreational boating also would be constrained to the smallest vessels. The majority of marine dredging operations (and the disposal of the dredged material) will require that appropriate licences are obtained from the relevant regulatory authorities, and dredging is usually carried out by (or for) harbour companies or corresponding government agencies.

Types of Dredging Vessels

Suction

The dredge drag head of a suction dredge barge on the Vistula River, Warsaw, Poland

The Geopotes 14 lifting its boom on a canal in The Netherlands. (*gēopotēs* is Greek for "that which drinks earth")

These operate by sucking through a long tube, like some vacuum cleaners but on a larger scale.

A plain suction dredger has no tool at the end of the suction pipe to disturb the material. This is often the most commonly used form of dredging.

Trailing Suction

A trailing suction hopper dredger (TSHD) trails its suction pipe when working. The pipe, which is fitted with a dredge drag head, loads the dredge spoil into one or more hoppers in the vessel. When the hoppers are full, the TSHD sails to a disposal area and either dumps the material through doors in the hull or pumps the material out of the hoppers. Some dredges also self-offload using drag buckets and conveyors.

The largest trailing suction hopper dredgers in the world are currently Jan De Nul's *Cristobal Colon* (launched 4 July 2008) and its sister ship *Leiv Eriksson* (launched 4 September 2009). Main design specs for the *Cristobal Colon* and the *Leiv Eriksson* are: 46,000 cubic metre hopper and a design dredging depth of 155 m. Next largest is *HAM 318* (Van Oord) with its 37,293 cubic metre hopper and a maximum dredging depth of 101 m.

Cutter-suction

A cutter-suction dredger's (CSD) suction tube has a cutting mechanism at the suction inlet. The cutting mechanism loosens the bed material and transports it to the suction mouth. The dredged material is usually sucked up by a wear-resistant centrifugal pump and discharged either through a pipe line or to a barge. Cutter-suction dredgers are most often used in geological areas consisting of hard surface materials (for example gravel deposits or surface bedrock) where a standard suction dredger would be ineffective. In recent years, dredgers with more powerful cutters have been built in order to excavate harder rock without the need for blasting.

The two largest cutter suction dredgers in the world are currently (as at August 2009) DEME's *D'Artagnan* (28,200 kW total installed power) and Jan De Nul's *J.F.J. DeNul* (27,240 kW). both built by IHC Merwede.

Auger Suction

This process functions like a cutter suction dredger, but the cutting tool is a rotating Archimedean screw set at right angles to the suction pipe. The first widely used auger dredges were designed in the 1980s by Mud Cat Dredges, which was run by National Car Rental, but is now a Division of Ellicott Dredges. In 1996, IMS Dredges introduced a self-propelled version of the auger dredge that allows the system to propel itself without the use of anchors or cables. During the 1980s and 1990s auger dredges were primarily used for sludge removal applications from waste water treatment plants. Today, auger dredges are used for a wider variety of applications including river maintenance and sand mining.

The most common auger dredge on the global market today is the Versi-Dredge. The turbidity shroud on auger dredge systems creates a strong suction vacuum, causing much less turbidity than conical (basket) type cutterheads and so they are preferred for environmental applications. The vacuum created by the shroud and the ability to convey material to the pump faster makes auger dredge systems more productive than similar sized conical (basket) type cutterhead dredges.

Jet-lift

These use the Venturi effect of a concentrated high-speed stream of water to pull the nearby water, together with bed material, into a pipe.

Air-lift

An airlift is a type of small suction dredge. It is sometimes used like other dredges. At other times, an airlift is handheld underwater by a diver. It works by blowing air into the pipe, and that air, being lighter than water, rises inside the pipe, dragging water with it.

Bucket

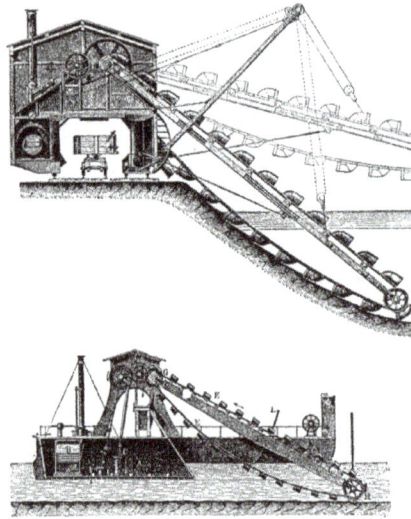

Bucket dredging

A bucket dredger is equipped with a bucket dredge, which is a device that picks up sediment by mechanical means, often with many circulating buckets attached to a wheel or chain. Some bucket dredgers and grab dredgers are powerful enough to rip out coral to make a shipping channel through coral reefs.

Clamshell

Clamshell dredging in process in Port Canaveral, Florida

A grab dredger picks up seabed material with a clam shell bucket, which hangs from an onboard crane or a crane barge, or is carried by a hydraulic arm, or is mounted like on a dragline. This technique is often used in excavation of bay mud. Most of these dredges are crane barges with spuds.

Backhoe/Dipper

A backhoe/dipper dredge has a backhoe like on some excavators. A crude but usable backhoe dredger can be made by mounting a land-type backhoe excavator on a pontoon. The six largest backhoe dredgers in the world are currently the Vitruvius, the Mimar Sinan, Postnik Jakovlev (Jan De Nul), the Samson (DEME), the Simson and the Goliath (Van Oord). They featured barge-mounted excavators. Small backhoe dredgers can be track-mounted and work from the bank of ditches. A backhoe dredger is equipped with a half-open shell. The shell is filled moving towards the machine. Usually dredged material is loaded in barges. This machine is mainly used in harbors and other shallow water.

Water Injection

A water injection dredger uses a small jet to inject water under low pressure (to prevent the sediment from exploding into the surrounding waters) into the seabed to bring the sediment in suspension, which then becomes a turbidity current, which flows away down slope, is moved by a second burst of water from the WID or is carried away in natural currents. Water injection results in a lot of sediment in the water which makes measurement with most hydrographic equipment (for instance: singlebeam echosounders) difficult.

Pneumatic

These dredgers use a chamber with inlets, out of which the water is pumped with the inlets closed. It is usually suspended from a crane on land or from a small pontoon or barge. Its effectiveness depends on depth pressure.

Bed Leveler

Steam dredger *Bertha*, built 1844, on a demonstration run in 1982

This is a bar or blade which is pulled over the seabed behind any suitable ship or boat. It has an effect similar to that of a bulldozer on land. The chain-operated steam dredger *Bertha*, built in 1844 to a design by Brunel and now the oldest operational steam vessel in Britain, was of this type.

Krabbelaar

This is an early type of dredger which was formerly used in shallow water in the Netherlands. It was a flat-bottomed boat with spikes sticking out of its bottom. As tide current pulled the boat, the spikes scraped seabed material loose, and the tide current washed the material away, hopefully to deeper water. *Krabbelaar* is Dutch for "scratcher".

Snagboat

A snagboat is designed to remove big debris such as dead trees and parts of trees from rivers and canals.

Amphibious

Some of these are any of the above types of dredger, which can operate normally, or by extending legs, also known as spuds, so it stands on the seabed with its hull out of the water. Some forms can go on land.

Some of these are land-type backhoe excavators whose wheels are on long hinged legs so it can drive into shallow water and keep its cab out of water. Some of these may not have a floatable hull and, if so, cannot work in deep water.

- Oliver Evans (1755–1819) in 1804 invented an amphibious dredger which was America's first steam-powered road vehicle.

Submersible

These are usually used to recover useful materials from the seabed. Many of them travel on continuous track. A unique variant is intended to walk on legs on the seabed.

Fishing

Dredge haul including live clams and empty shells

Fishing dredges are used to collect various species of clams scallops, oysters or crabs from the seabed. These dredges have the form of a scoop made of chain mesh, and are towed by a fishing boat. Careless dredging can be destructive to the seabed. Nowadays some scallop dredging is replaced by collecting via scuba diving.

Police Drag

In some police departments a small dredge (sometimes called a *drag*) is used to find and recover objects and bodies from underwater. The bodies may be murder victims, or people who committed suicide by drowning, or victims of accidents. It is sometimes pulled by people walking on the bank. Search and rescue units also often use this type of dredge in searching for bodies of missing persons.

Disposal of Materials

In a "hopper dredger", the dredged materials end up in a large onboard hold called a "hopper." A suction hopper dredger is usually used for maintenance dredging. A hopper dredge usually has doors in its bottom to empty the dredged materials, but some dredges empty their hoppers by splitting the two halves of their hulls on giant hydraulic hinges. Either way, as the vessel dredges, excess water in the dredged materials is spilled off as the heavier solids settle to the bottom of the hopper. This excess water is returned to the sea to reduce weight and increase the amount of solid material (or slurry) that can be carried in one load. When the hopper is filled with slurry, the dredger stops dredging and goes to a dump site and empties its hopper.

Some hopper dredges are designed so they can also be emptied from above using pumps if dump sites are unavailable or if the dredge material is contaminated. Sometimes the slurry of dredgings and water is pumped straight into pipes which deposit it on nearby land. Other times, it is pumped into barges (also called scows), which deposit it elsewhere while the dredge continues its work.

A number of vessels, notably in the UK and NW Europe de-water the hopper to dry the cargo to enable it to be discharged onto a quayside 'dry'. This is achieved principally using self discharge bucket wheel, drag scraper or excavator via conveyor systems.

When contaminated (toxic) sediments are to be removed, or large volume inland disposal sites are unavailable, dredge slurries are reduced to dry solids via a process known as dewatering. Current dewatering techniques employ either centrifuges, Geotube containers, large textile based filters or polymer flocculant/congealant based apparatus.

In many projects, slurry dewatering is performed in large inland settling pits, although this is becoming less and less common as mechanical dewatering techniques continue to improve.

Similarly, many groups (most notable in east Asia) are performing research towards utilizing dewatered sediments for the production of concretes and construction block, although the high organic content (in many cases) of this material is a hindrance toward such ends.

Environmental Impacts

Dredging can create disturbance to aquatic ecosystems, often with adverse impacts. In addition, dredge spoils may contain toxic chemicals that may have an adverse effect on the disposal area; furthermore, the process of dredging often dislodges chemicals residing in benthic substrates and injects them into the water column.

The activity of dredging can create the following principal impacts to the environment:

- Release of toxic chemicals (including heavy metals and PCB) from bottom sediments into the water column.

- Collection of heavy metals lead left by fishing, bullets, 98% mercury reclaimed [natural occurring and left over from gold rush era].

- Short term increases in turbidity, which can affect aquatic species metabolism and interfere with spawning. Suction dredging activity is allowed only during non-spawing time frames set by fish and game (in-water work periods).

- Secondary impacts to marsh productivity from sedimentation

- Tertiary impacts to avifauna which may prey upon contaminated aquatic organisms

- Secondary impacts to aquatic and benthic organisms' metabolism and mortality

- Possible contamination of dredge spoils sites

- Changes to the topography by the creation of "spoil islands" from the accumulated spoil

- Releases toxic compound Tributyltin, a popular biocide used in anti-fouling paint banned in 2008, back into the water.

The nature of dredging operations and possible environmental impacts cause the industry to be closely regulated and a requirement for comprehensive regional environmental impact assessments with continuous monitoring. The U.S. Clean Water Act requires that any discharge of dredged or fill materials into "waters of the United States," including wetlands, is forbidden unless authorized by a permit issued by the Army Corps of Engineers. As a result of the potential impacts to the environment, dredging is restricted to licensed areas only with vessel activity monitored closely using automatic GPS systems.

Major Dredging Companies

According to a Rabobank outlook report in 2013, the largest dredging companies in the world are in order of size:

- Van Oord Dredging and Marine Contractors (Netherlands)

- China Harbour Engineering (China)

- Jan De Nul (Belgium)

- DEME (Belgium)

- Qingzhou Hengchuan Ore Machinery Co.,Ltd (China)

- Royal Boskalis Westminster(Netherlands)

Dredge Monitoring Software

Dredgers are often equipped with dredge monitoring software to help the dredge operator position the dredger and monitor the current dredge level. The monitoring software often uses Real Time Kinematic satellite navigation to accurately record where the machine has been operating and to what depth the machine has dredged to.

References

- Montrie, Chad (2003). To Save the Land and People: A History of Opposition to Surface Coal Mining in Appalachia. United States: The University of North Carolina Press. p. 17. ISBN 0-8078-2765-7.

- Mondal, Sudipto (17 March 2015). "IAS officer who took on sand mafia found dead in Bengaluru residence". Hindustan Times. Retrieved 17 March 2015.

- Burns, Shirley Stewart (2005). "Bringing Down the Mountains: the Impact of Mountaintop Removal Surface Coal Mining on Southern West Virginia Communities, 1970–2004" (PDF). Ph.D. dissertation. West Virginia University. Retrieved 2013-03-25.

- "National Institute of Oceanography, India". Web.archive.org. Archived from the original on 12 January 2009. Retrieved 14 June 2013.

- Walker, Margaret (1991). "What price Tasmanian scallops? A report of morbidity and mortality associated with the scallop diving season in Tasmania 1990.". South Pacific Underwater Medicine Society Journal. 21 (1). Retrieved 16 July 2013.

- "Jan de Nul's mega trailer Cristóbal Colón launched - Dredging News Online". Sandandgravel.com. 7 July 2008. Retrieved 14 June 2013.

- "Keel-laying ceremony for Jan de Nul's Leiv *Eiriksson held - Dredging News Online". Sandandgravel.com. 1 September 2008. Retrieved 14 June 2013.*

Understanding Underground Mining

Underground mining can best be understood in regard to drift mining, stoping, shaft mining and drilling and blasting and other techniques. Stoping is the extraction of the desired ore or any other mineral from an underground mine, while shaft mining is excavating a tunnel from the top down where there is initially no access to the bottom. This chapter develops a profound understanding in the reader, related to underground mining.

Underground Mining (Hard Rock)

Underground hard rock mining refers to various underground mining techniques used to excavate *hard* minerals, mainly those containing metals such as ore containing gold, silver, iron, copper, zinc, nickel, tin and lead, but also involves using the same techniques for excavating ores of gems such as diamonds. In contrast soft rock mining refers to excavation of softer minerals such as salt, coal, or oil sands.

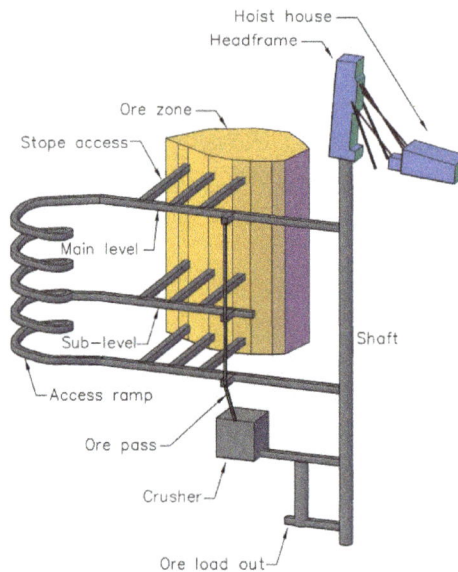

A three-dimensional model of an underground mine with shaft access

Mine Access

Underground Access

Accessing underground ore can be achieved via a decline (ramp), inclined vertical shaft or adit.

Decline portal

- Declines can be a spiral tunnel which circles either the flank of the deposit or circles around the deposit. The decline begins with a box cut, which is the portal to the surface. Depending on the amount of overburden and quality of bedrock, a galvanized steel culvert may be required for safety purposes. They may also be started into the wall of an open cut mine.

- Shafts are vertical excavations sunk adjacent to an ore body. Shafts are sunk for ore bodies where haulage to surface via truck is not economical. Shaft haulage is more economical than truck haulage at depth, and a mine may have both a decline and a ramp.

- Adits are horizontal excavations into the side of a hill or mountain. They are used for horizontal or near-horizontal ore bodies where there is no need for a ramp or shaft.

Declines are often started from the side of the high wall of an open cut mine when the ore body is of a payable grade sufficient to support an underground mining operation but the strip ratio has become too great to support open cast extraction methods. They are also often built and maintained as an emergency safety access from the underground workings and a means of moving large equipment to the workings.

Ore Access

Levels are excavated horizontally off the decline or shaft to access the ore body. Stopes are then excavated perpendicular (or near perpendicular) to the level into the ore.

Development mining vs. production mining

There are two principal phases of underground mining: development mining and production mining.

Development mining is composed of excavation almost entirely in (non-valuable) waste rock in order to gain access to the orebody. There are six steps in development mining: remove previously blasted material (muck out round), Scaling (removing any unstable slabs of rock hanging from the roof and sidewalls to protect workers and equipment from damage), installing support or/ and reinforcement using shotcrete etceteras, drill face rock, load explosives, and blast explosives. To start the mining first step is to make the path to go down. The path is defined as 'Decline' as describe above. Before the start of Decline all preplanning of Power facility, drilling arrangement, dewatering, ventilation and, muck withdrawal facilities are required.

Production mining is further broken down into two methods, long hole and short hole. Short hole mining is similar to development mining, except that it occurs in ore. There are several different methods of long hole mining. Typically long hole mining requires two excavations within the ore at different elevations below surface, (15 m – 30 m apart). Holes are drilled between the two excavations and loaded with explosives. The holes are blasted and the ore is removed from the bottom excavation.

Ventilation

Door for directing ventilation in an old lead mine. The ore hopper at the front is not part of the ventilation.

One of the most important aspects of underground hard rock mining is ventilation. Ventilation is the primary method of clearing hazardous gases and/or dust which are created from drilling and blasting activity (e.g., silica dust, NOx), diesel equipment (e.g., diesel particulate, carbon monoxide), or to protect against gases that are naturally emanating from the rock (e.g., radon gas). Ventilation is also used to manage underground temperatures for the workers. In deep, hot mines ventilation is used to cool the workplace; however, in very cold locations the air is heated to just above freezing before it enters the mine. Ventilation raises are typically used to transfer ventilation from surface to the workplaces, and can be modified for use as emergency escape routes. The primary sources of heat in underground hard rock mines are virgin rock temperature, machinery, auto compression, and fissure water. Other small contributing factors are human body heat and blasting.

Ground Support

Some means of support is required in order to maintain the stability of the openings that are excavated. This support comes in two forms, local support and area support.

Area Ground Support

Area ground support is used to prevent major ground failure. Holes are drilled into the back (ceiling) and walls and a long steel rod (or rock bolt) is installed to hold the ground together. There are three categories of rock bolt, differentiated by how they engage the host rock. They are:

Mechanical Bolts

- Point anchor bolts (or expansion shell bolts) are a common style of area ground support. A point anchor bolt is a metal bar between 20 mm – 25 mm in diameter, and between 1 m – 4 m long (the size is determined by the mine's engineering department). There is an expansion shell at the end of the bolt which is inserted into the hole. As the bolt is tightened by the installation drill the expansion shell expands and the bolt tightens holding the rock together. Mechanical bolts are considered temporary support as their lifespan is reduced by corrosion as they are not grouted.

Grouted Bolts

- Resin grouted rebar is used in areas which require more support than a point anchor bolt can give. The rebar used is of similar size as a point anchor bolt but does not have an expansion shell. Once the hole for the rebar is drilled, cartridges of polyester resin are installed in the hole. The rebar bolt is installed after the resin and spun by the installation drill. This opens the resin cartridge and mixes it. Once the resin hardens the drill spinning tightens the rebar bolt holding the rock together. Resin grouted rebar is considered a permanent ground support with a lifespan of 20–30 years.

- Cable bolts are used to bind large masses of rock in the hanging wall and around large excavations. Cable bolts are much larger than standard rock bolts and rebar, usually between 10–25 metres long. Cable bolts are grouted with a cement grout.

Friction Bolts

- Friction stabilizer (frequently called by the genericized trademark *Split Set*) are much easier to install than mechanical bolts or grouted bolts. The bolt is hammered into the drill hole, which has a smaller diameter than the bolt. Pressure from the bolt on the wall holds the rock together. Friction stabilizers are particularly susceptible to corrosion and rust from water unless they are grouted. Once grouted the friction increases by a factor of 3-4.

- Swellex is similar to Friction stabilizers, except the bolt diameter is smaller than the hole diameter. High pressure water is injected into the bolt to expand the bolt diameter to hold the rock together. Like the friction stabilizer, swellex is poorly protected from corrosion and rust.

Local Ground Support

Local ground support is used to prevent smaller rocks from falling from the back and ribs. Not all excavations require local ground support.

- Welded Wire Mesh is a metal screen with 10 cm x 10 cm (4 inch) openings. It is held to the back using point anchor bolts or resin grouted rebar.

- Shotcrete is fibre reinforced spray on concrete which coats the back and ribs preventing smaller rocks from falling. Shotcrete thickness can be between 50 mm – 100 mm.

- Latex Membranes can be sprayed on the backs and ribs similar to shotcrete, but in smaller amounts.

Stope and Retreat vs. Stope and Fill

Stope and Retreat

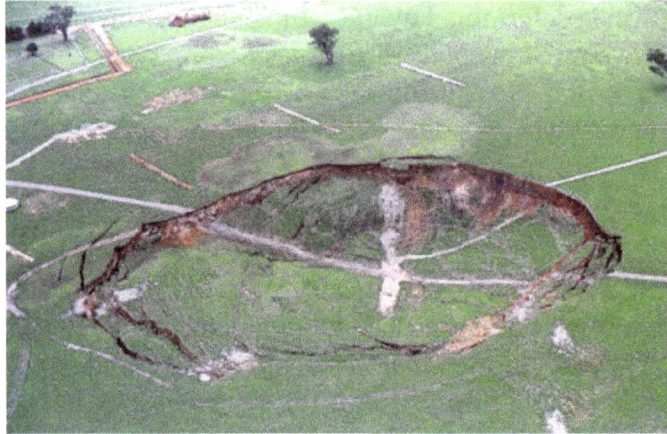

Sub-Level Caving Subsidence reaches surface at the Ridgeway underground mine.

Using this method, mining is planned to extract rock from the stopes without filling the voids; this allows the wall rocks to cave in to the extracted stope after all the ore has been removed. The stope is then sealed to prevent access.

Stope and Fill

Where large bulk ore bodies are to be mined at great depth, or where leaving pillars of ore is uneconomical, the open stope is filled with backfill, which can be a cement and rock mixture, a cement and sand mixture or a cement and tailings mixture. This method is popular as the refilled stopes provide support for the adjacent stopes, allowing total extraction of economic resources.

Mining Methods

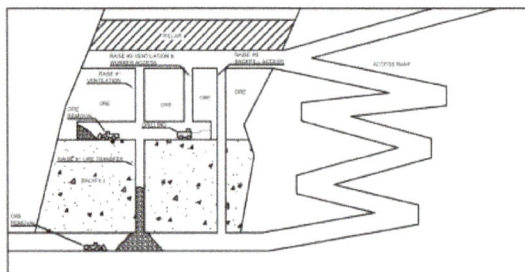

Schematic diagram of cut and fill mining

The mining method selected is determined by the size, shape, orientation and type of orebody to be mined. The orebody can be narrow vein such as a gold mine in the Witwatersrand, the orebody can be massive similar to the Olympic Dam mine, South Australia, or Cadia-Ridgeway Mine, New South Wales. The width or size of the orebody is determined by the grade as well as the distribution of the ore. The dip of the orebody also has an influence on the mining method for example a narrow horizontal vein orebody will be mined by room and pillar or a longwall method whereas a vertical narrow vein orebody will be mined by an open stoping or cut and fill method. Further consideration is needed for the strength of the ore as well as the surrounding rock. An orebody hosted in strong self-supporting rock may be mined by an open stoping method and an orebody hosted in poor rock may need to be mined by a cut and fill method where the void is continuously filled as the ore is removed.

Selective Mining Methods

- Cut and fill mining is a method of short-hole mining used in steeply dipping or irregular ore zones, in particular where the hanging wall limits the use of long-hole methods. The ore is mined in horizontal or slightly inclined slices, and then filled with waste rock, sand or tailings. Either fill option may be consolidated with concrete, or left unconsolidated. Cut and fill mining is an expensive but selective method, with low ore loss and dilution.

- Drift and fill is similar to cut and fill, except it is used in ore zones which are wider than the method of drifting will allow to be mined. In this case the first drift is developed in the ore, and is backfilled using consolidated fill. The second drift is driven adjacent to the first drift. This carries on until the ore zone is mined out to its full width, at which time the second cut is started atop of the first cut.

- Shrinkage stoping is a short-hole mining method which is suitable for steeply dipping orebodies. The method is similar to cut and fill mining with the exception that after being blasted, broken ore is left in the stope where it is used to support the surrounding rock and as a platform from which to work. Only enough ore is removed from the stope to allow for drilling and blasting the next slice. The stope is emptied when all of the ore has been blasted. Although it is very selective and allows for low dilution, since most of the ore stays in the stope until mining is completed there is a delayed return on capital investments.

- Room and pillar mining : Room and pillar mining is commonly done in flat or gently dipping bedded ore bodies. Pillars are left in place in a regular pattern while the rooms are mined out. In many room and pillar mines, the pillars are taken out starting at the farthest point from the stope access, allowing the roof to collapse and fill in the stope. This allows for greater recovery as less ore is left behind in pillars.

- VRM: Vertical retrieval mining is a method where mine is decided in vertical zones with depth of about 50 meters. Long-hole high diamond drilling is done through ITH drills. Material retrieval is done from bottom of the section developed. Ore blasted in retrieval taken in phase. Last cleaning of ore is done through remote controlled LHD machines. The zone is now back filled using cemented mix fill. Side chambers will be mined in pre-planned sequence after the fill has solidified.

Bulk Mining Methods

- Block caving is used to mine massive steeply dipping orebodies (typically low grade) with high friability. An undercut with haulage access is driven under the orebody, with "drawbells" excavated between the top of the haulage level and the bottom of the undercut. The drawbells serve as a place for caving rock to fall into. The orebody is drilled and blasted above the undercut, and the ore is removed via the haulage access. Due to the friability of the orebody the ore above the first blast caves and falls into the drawbells. As ore is removed from the drawbells the orebody caves in, providing a steady stream of ore. If caving stops and removal of ore from the drawbells continues, a large void may form, resulting in the potential for a sudden and massive collapse and potentially catastrophic windblast throughout the mine. Where caving does continue, the ground surface may collapse into a surface depression such as those at the Climax and Henderson molybdenum mines in Colorado. Such a configuration is one of several to which miners apply the term "glory hole".

Orebodies that do not cave readily are sometimes preconditioned by hydraulic fracturing, blasting, or by a combination of both. Hydraulic fracturing has been applied to preconditioning strong roof rock over coal longwall panels, and to inducing caving in both coal and hard rock mines.

Ore Removal

In mines which use rubber tired equipment for coarse ore removal, the ore (or "muck") is removed from the stope (referred to as "mucked out" or "bogged") using center articulated vehicles (referred to as boggers or LHD (Load, Haul, Dump machine)). These pieces of equipment may operate using diesel engines or electric motors, and resemble a low-profile front end loader. LHD operated through electricity utilize trailing cables which are flexible and can be extended or retracted on a reel.

The ore is then dumped into a truck to be hauled to the surface (in shallower mines). In deeper mines the ore is dumped down an ore pass (a vertical or near vertical excavation) where it falls to a collection level. On the collection level, it may receive primary crushing via jaw or cone crusher, or via a rockbreaker. The ore is then moved by conveyor belts, trucks or occasionally trains to the shaft to be hoisted to the surface in buckets or skips and emptied into bins beneath the surface headframe for transport to the mill.

In some cases the underground primary crusher feeds an inclined conveyor belt which delivers ore via an incline shaft direct to the surface. The ore is fed down ore passes, with mining equipment accessing the ore body via a decline from surface.

Underground Mining (Soft Rock)

Underground mining (soft rock) refers to a group of underground mining techniques used to extract coal, oil shale, potash and other minerals or geological materials from sedimentary ("soft") rocks. Because deposits in sedimentary rocks are commonly layered and relatively less hard, the mining methods used differ from those used to mine deposits in igneous or metamorphic rocks (see Underground mining (hard rock)). Underground mining techniques also differ greatly from those of surface mining.

Methods

- Longwall mining – A set of longwall mining equipment consists of a coal shearer mounted on conveyor operating underneath a series of self-advancing hydraulic roof supports. Almost the entire process can be automated. Longwall mining machines are typically 150–250 metres in width and 1.5 to 3 metres high. Longwall miners extract "panels" - rectangular blocks of coal as wide as the face the equipment is installed in, and as long as several kilometres. Powerful mechanical coal cutters (shearers) cut coal from the face, which falls onto an armoured face conveyor for removal. Longwalls can advance into an area of coal, or more commonly, retreat back between development tunnels (called "gateroads") As a longwall miner retreats back along a panel, the roof behind the supports is allowed to collapse in a planned and controlled manner.

- Room-and-pillar mining or continuous mining – Room and pillar mining is commonly done in flat or gently dipping bedded ores. Pillars are left in place in a regular pattern while the rooms are mined out. In many room and pillar mines, the pillars are taken out, starting at the farthest point from the mine haulage exit, retreating, and letting the roof come down upon the floor. Room and pillar methods are well adapted to mechanization, and are used in deposits such as coal, potash, phosphate, salt, oil shale, and bedded uranium ores.

- Blast mining – An older practice of coal mining that uses explosives such as dynamite to break up the coal seam, after which the coal is gathered and loaded onto shuttle cars or conveyors for removal to a central loading area. This process consists of a series of operations that begins with "cutting" the coalbed so it will break easily when blasted with explosives. This type of mining accounts for less than 5% of total underground production in the U.S. today.

- Shortwall mining – A coal mining method that accounts for less than 1% of deep coal production, shortwall involves the use of a continuous mining machine with moveable roof supports, similar to longwall. The continuous miner shears coal panels 150–200 feet wide and more than a half-mile long, depending on other things like the strata of the Earth and the transverse waves.

- Coal skimming – While no longer in general use, because of the massive amount of water needed and environmental damage thereof, in the late 1930s DuPont developed a method that was much faster and less labour-intensive than previous methods to separate the lighter coal from the mining refuse (e.g. slate) called "coal skimming" or the "sink and float method".

Mine Shorthand

The number sign, or hash sign (#) is often used as shorthand to denote shaft or seam, as in 4# (4 shaft or seam depending on context).

Drift Mining

Drift mining is either the mining of an ore deposit by underground methods, or the working of coal seams accessed by adits driven into the surface outcrop of the coal bed. A drift mine is an underground mine in which the entry or access is above water level and generally on the slope of a hill, driven horizontally into the ore seam. Random House dictionary says the origin of the term "drift mine" is an Americanism, circa 1885-1890.

Drift is a more general mining term, meaning a near-horizontal passageway in a mine, following the bed (of coal, for instance) or vein of ore. A drift may or may not intersect the ground surface. A Drift is a horizontal passage underground. A drift follows the vein, as distinguished from a crosscut that intersects it, or a level or gallery, which may do either. All horizontal or subhorizontal development openings made in a mine have the generic name of drift. These are simply tunnels made in the rock, with a size and shape depending on their use—for example, haulage, ventilation, or exploration.

Historical US Drift Mining (Coal)

Coal miner standing in a drift portal at Fork Mountain, TN, 1920.

This section provides a very abbreviated snapshot of the drift mining information generally available.

Illinois: Argyle Lake State Park's website says the Argyle Hollow (occupied by a lake since 1948) region has been rich in coal, clay and limestone resources. Historically, individuals commonly opened and dug their own "drift mines" to supplement their income. In Appalachia, small coal mining operations such as these were known as "country bank" or "farmer" coal mines, and usually produced only small quantities for local use.

Indiana: The Lusk Mine, now in Turkey Run State Park, was in operation from the late 1800s through the late 1920s. Too small for commercial operation, the mine probably provided coal for the Lusk family and later for the park.

Kentucky: In 1820 the first commercial mine in Kentucky, known as the "McLean drift bank" opened near the Green River and Paradise in Muhlenberg County. In Drift, KY, Beaver Coal & Mining Company was the most well known operator of mines, but there were other smaller mines (Floyd-Elkhorn Consolidated Collieries, Turner-Elkhorn Coal Company, etc.) as well.

Maryland: Dorsey Coal Company's Ashby coal mine, a small drift mine probably in the Upper Freeport coal; and the mine of the Taylor-Offutt Coal Company near Oakland, MD.

Ohio: In the 1880s, State Inspector of Mines, Andrew Roy, issued a report on the The Mines and Mining Resources of Ohio, which includes the following paragraphs:

The capacity or output of the mines of the State varies greatly. Thick coals are capable of a greater daily output than thin seams, and as a general rule drift mines possess greater advantages for loading coal rapidly than shaft openings. In many of the mines of the great vein region of the Hocking valley the capacity is equal to 1,200 to 1,500 tons per day. In shaft mines 600 to 700 tons daily is regarded as a good output.

The first ton of coal in a shaft mine 100 feet in depth and having a daily capacity of 600 tons frequently costs the mining adventurer upwards of $20,000 (1888), and cases are on record where owing to the extraordinary amount of water in sinking, $100,000 (1888) have been expended before coal was reached. Drift mines, as they require no machinery for pumping water and raising coal, cost less than half the amount required in shaft mining.

Water is, however, an expensive item in drift mines opened on the dip of the coal, and underground hauling under such conditions is unusually costly.

Drift mine entry in West Virginia, 1908. Photo by Lewis Hine.

Pennsylvania: Indiana County: Graff Drift Mines, near Blairsville. Commodore Mines, Nos. 1 & 3 (drift mines), No. 2 (slope mine), Green Twp. Empire "F" Mine (1910-?), Shanktown; a drift mine, mining the "B" coal seam, mining done by machine, owner Pioneer Coal Company, Clearfield. Empire "M" Mine (McKean Mine) (1906-?), a drift mine, non-gaseous, mining a 38" thick seam of Lower Kittaning coal using compressed air machines; ventilation provided by a 8' Stine steam-powered fan, Clymer, PA. Rodkey Mine(1906-?), a drift mine, Clymer. Ernest Mine No. 2 (1903-1965), a drift mine, at Ernest, Rayne Twp., Indiana Co., PA.

Tennessee: The Fork Mountain, TN, drift portal entered an 84-inch unnamed seam of coal. Most coal seams in Tennessee were not this thick.

West Virginia: Many, many references to and photographs of WV drift mines in the Scrapbook of

Appalachian Coal Towns, including Sprague, Kaymoor, Nuttallburg, Venus, Layland, Elverton, Casselman (aka Castleman), etc.

Historical US Drift Mining (Gold)

Alaska: Drift mining methods were used extensively to mine placer deposits during the early years (1899-) of the Nome mining district. During summer, surface deposits could be worked, but some placer deposits were buried too deeply for surface placering. In addition, water to wash the gold from the placers was not available in the winter. Many miners tunneled into deep placer deposits, bringing out the high-grade gravels to be washed at the spring thaw. Most of the ground in Nome is permafrost. By drift mining, miners were able to recover much of the gold buried under the permafrost.

Gold at Nome was concentrated in three ancient beach lines, now inshore, above sea level, and buried under roughly fifty feet of permafrost overlain by two feet of tundra. Gold was usually found on top of "false bedrock," a layer of clay that occurred at the base of the beach or stream deposit. Miners initially sank shafts to prospect for the pay streaks by building a fire atop the permafrost, then as it melted, shoveling away the mud. The process would continue down to either a pay streak or bedrock.

When gold was found drift mining began. Miners would tunnel horizontally from the bottom of their prospect shaft to follow the gold along the surface of the bedrock. The tunnels did not cave in because the ground was frozen. Miners discovered old underground beach and river gravels rich with gold. Around 1900 the population of Nome was more than twenty thousand, many of them drift miners. Nome's gold fields, appearing untouched from the surface, are honeycombed with tunnels left by the gold rush drift miners. Today's miners, prospecting with modern drilling equipment, sometimes hit old drifts; this was, and is, a technique copied from the Welsh coal miners of south Wales and is much more effective than using bell pits.

California: Gold has been mined from placer gold deposits up and down the state and in different types of environment. Initially, rich, easily discovered, surface and river placers were mined until about 1864. Hydraulic mines, using powerful water cannons to wash whole hillsides, were the chief sources of gold for the next 20 years. In 1884, Judge Lorenzo Sawyer issued a decree prohibiting the dumping of hydraulic mining debris into the Sacramento River, effectively eliminating large-scale hydraulic operations. For the next 14 years, drift mining placer gold deposits in buried Tertiary channels partially made up for the loss of placer gold production, but overall production declined. Production rose again with the advent of large-scale dredging. The first successful gold dredge was introduced on the lower Feather River near Oroville in 1898.

Drift Mines in Current Production

Safety and Environmental Issues

Drift mines in eastern Kentucky are subject to roof collapse due to hillseams, especially within 100 feet of the portal.

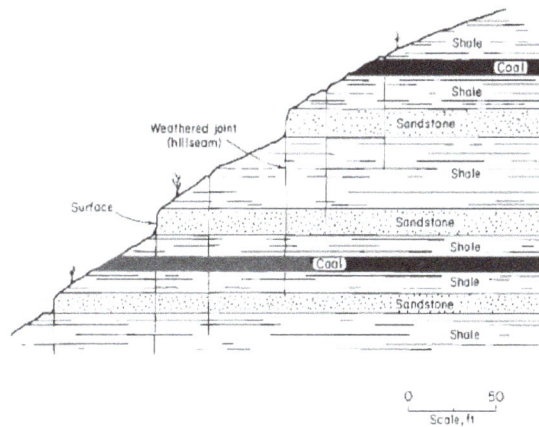

Profile of hillseam occurrence.

Profile of hillseam occurrence

In 1989 the U.S. Bureau of Mines published a study of eastern Kentucky drift mines as part of an ongoing research program to characterize the outcrop barrier zone. "Hillseams" were identified as the dominant geologic cause of roof instability unique to the outcrop barrier zone, with many roof fall injuries and fatalities attributed to them. Hillseam is the eastern Kentucky miners term for weather-enlarged tension joints that occur in shallow mine overburden where surface slopes are steep. Hillseams are most conspicuous within 200 ft laterally of a coalbed outcrop and under 300 ft or less of overburden. Hillseams form by stress relief, and therefore tend to parallel topographic contours and ridges. They can intersect at various angles, especially under the nose of a ridge, and create massive blocks or wedges of roof prone to failure. Examples of hillseams are described in both outcrop and in coal mine roof to establish their geologic character and contribution to roof failure.

Stoping

Sketch painting of miners stoping at the Burra Burra Mine, Burra, Australia, 1847.

Stoping with an air drill in an American iron mine in the 20th century (museum exhibit)

Stoping is the process of extracting the desired ore or other mineral from an underground mine, leaving behind an open space known as a stope. Stoping is used when the country rock is sufficiently strong not to collapse into the stope, although in most cases artificial support is also provided.

The earliest forms of stoping were conducted with hand tools or by fire-setting; later gunpowder was introduced. From the 19th century onward, various other explosives, power-tools, and machines came into use. As mining progresses the stope is often backfilled with tailings, or when needed for strength, a mixture of tailings and cement. In old mines, stopes frequently collapse at a later time, leaving craters at the surface. They are an unexpected danger when records of underground mining have been lost with the passage of time.

Stoping is considered "productive work", and is contrasted with "deadwork", the work required merely to access the mineral deposit, such as sinking shafts and winzes, carving adits, tunnels, and levels, and establishing ventilation and transportation.

Overview

A stope can be created in a variety of ways. The specific method of stoping depends on a number of considerations, both technical and economical, based largely on the geology of the ore body being mined. These include the incline of the deposit (whether it is flat, tilted or vertical), the width of the deposit, the grade of the ore, the hardness and strength of the surrounding rock, and the cost of materials for supports.

It is common to dig shafts vertically downwards to reach the ore body and then drive horizontal levels through it. Stoping then takes place from these levels.

When the ore body is more or less horizontal, various forms of room and pillar stoping, cut and fill, or longwall mining can take place. In steeply-dipping ore bodies, such as lodes of tin, the stopes become long narrow near-vertical spaces, which, if one reaches the surface is known as a gunnis or coffen. A common method of mining such vertical ore bodies is stull stoping.

Open-stope Systems

A large stope in a salt mine in Poland - now converted into a tourist attraction

Open stoping is generally divided into two basic forms based on direction: overhand and underhand stoping, which refer to the removal of ore from above or below the level, respectively. It is also possible to combine the two in a single operation.

Underhand Stoping

Underhand stoping, also known as horizontal-cut underhand or underbreaking stoping, is the working of an ore deposit from the top downwards. Like shrinkage stoping, underhand stoping is most suitable for steeply dipping ore bodies. Because of the mechanical advantage it offers hand tools being struck downward (rather than upward, against gravity), this method was dominant prior to the invention of rock blasting and powered tools.

Overhand Stoping

In overhand stoping, the deposit is worked from the bottom upward, the reverse of underhand stoping. With the advent of rock blasting and power drills, it became the predominant direction of stoping.

Combined Stoping

In combined stoping, the deposit is simultaneously worked from the bottom upward and the top downward, combining the techniques of overhand and underhand stoping into a single approach.

Breast Stoping

Breast stoping is a method used in horizontal or near-horizontal ore bodies, where gravity is not usable to move the ore around. Breast stoping lacks the characteristic "steps" of either underhand or overhand stoping, being mined in a singular cut. Room and pillar is a type of breast stoping.

Timbered-stope Systems

Stull Stoping

Stull stoping is a form of stoping used in hardrock mining that uses systematic or random timbering ("stulls") placed between the foot and hanging wall of the vein. The method requires that the hanging wall and often the footwall be of competent rock as the stulls provide the only artificial support. This type of stope has been used up to a depth of 3,500 feet (1,077 m) and at intervals up to 12 feet (3.7 m) wide. The 1893 mining disaster at Dolcoath mine in Cornwall was caused by failure of the stulls holding up a huge weight of waste rock.

Square-set Stoping

Square-set stoping is a method relying on square-set timbering. Square set timbers are set into place as support and are then filled with cement. The cement commonly uses fine tailings. This is a highly specialized method of stoping requiring expert input. Square set was invented in the Comstock Lode, Virginia City, Nevada in the 1860s, and they utilized the waste rock to fill the stopes..

Shrinkage Stoping

A large stope in the Treadwell gold mine, ERWIN TAN JR; an example of shrinkage stoping

Shrinkage stoping, is most suitable for steeply dipping ore bodies (70°—90°). In shrinkage stoping, mining proceeds from the bottom upwards, in horizontal slices (similar to cut and fill mining), with the broken ore being left in place for miners to work from. Because blasted rock takes up a greater volume than in situ rock (due to swell factor), some of the blasted ore (approximately 40%) must be removed to provide working space for the next ore slice. Once the top of the stope is reached all the ore is removed from the stope. The stope may be backfilled or left empty, depending on the rock conditions.

Long Hole Stoping

Introduction

Long hole stoping as the name suggests uses holes drilled by a production drill to a predetermined pattern as designed by a Mining Engineer. Long hole stoping is a highly selective and productive method of mining and can cater for varying ore thicknesses and dips (0 - 90 degree). It differs from manual methods such as timbered and shrinkage as once the stope has begun blasting phase it cannot be accessed by personnel. For this reason the blasted rock is designed to fall into a supported drawpoint or removed with Tele Remote Control LHD. The biggest limitation with this method is the length of holes that can be accurately drilled by the production drill, larger diameter holes using ITH (In the Hole Hammer) drills can be accurate to over 100m in length while floating boom top hammer rigs are limited to ~30m.

Slot - Initial Void

Holes drilled underground are generally drilled perpendicular, in a radial pattern around the drive. For the blastholes to successfully extract the ore material they must be able to fire into a void in front. A slot is required in every stope to provide the initial void. The slot is often the most difficult, costly and highest risk component of mining a stope. Depending on the shape, height and other factors, different methods to create a slot can be used such as:

- Raise bore, a circular shaft mined bottom up using mechanical rollers to achieve shaft profile. This method works well in larger stopes however requires both access to top and bottom of stoping block, raise bore's work most effectively between 45 and 90 deg.

- Longhole rise, a pattern of tightly spaced blastholes and reamers (empty holes with no charge), similar to a burn cut in a development round. Can be done as downhole and fired in multiple lifts (15m rise in 3 lifts of 5m to minimise chance of blast failing) or as uphole in one single firing. This method works well for shorter raises between 45 - 90 deg, however is prone to freezing and remedial drilling is possibly required to extract slot to full height.

- Airleg Raise, using an airleg (jackleg) machine to develop a sub vertical raise into the stoping block. This method has the advantage of giving geological and geotechnical teams further analysis of the stoping block prior to mining.

- Boxhole Boring, similar to raise boring however less productive as broken material is extracted from the same location as the drill, is used to bore vertically with no top level access required.

Room and Pillar

Room and pillar (variant of breast stoping), also called pillar and stall, is a mining system in which the mined material is extracted across a horizontal plane, creating horizontal arrays of rooms and pillars. The ore is extracted in two phases. In the first, "pillars" of untouched material are left to support the roof overburden, and open areas or "rooms" are extracted underground; the pillars are then partially extracted in the same manner as in the "Bord & Pillar method". The technique is usually used for relatively flat-lying deposits, such as those that follow a particular stratum.

The room and pillar system is used in mining coal, iron and base metals ores, particularly when found as manto or blanket deposits, stone and aggregates, talc, soda ash and potash.

The key to successful room and pillar mining is in the selection of the optimum pillar size. In general practice, the size of both room and pillars are kept almost equal, while in Bord & Pillar, pillar size is much larger than bord (gallery). If the pillars are too small the mine will collapse, but if they are too large then significant quantities of valuable material will be left behind, reducing the profitability of the mine. The percentage of material mined varies depending on many factors, including the material mined, height of the pillar, and roof conditions; typical values are: stone and aggregates 75 percent, coal 60 percent, and potash 50 percent.

History

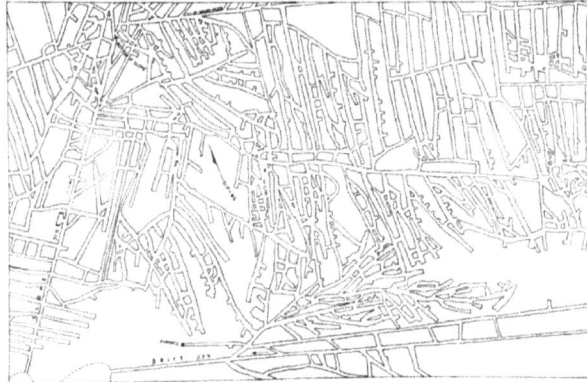

A Maryland coal mine from 1850

Room and pillar mining is one of the oldest mining methods. Early room and pillar mines were developed more or less at random, with pillar sizes determined empirically and headings driven in whichever direction was convenient.

Random mine layout makes ventilation planning difficult, and if the pillars are too small, there is the risk of pillar failure. In coal mines, pillar failures are known as squeezes because the roof squeezes down, crushing the pillars. Once one pillar fails, the weight on the adjacent pillars increases, and the result is a chain reaction of pillar failures. Once started, such chain reactions can be extremely difficult to stop, even if they spread slowly.

Mine Layout

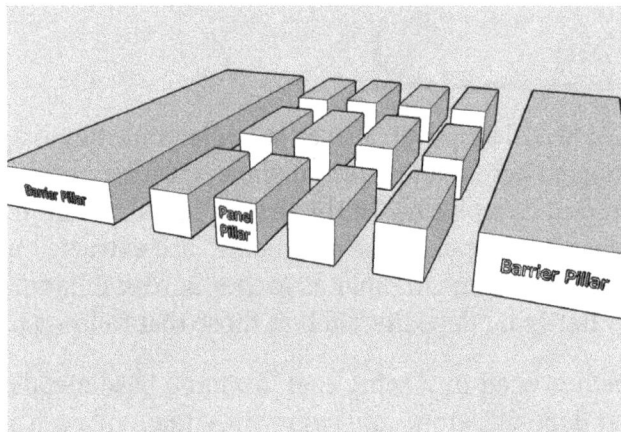

Room and pillar mines are developed on a grid basis except where geological features such as faults require the regular pattern to be modified. The size of the pillars is determined by calculation. The load-bearing capacity of the material above and below the material being mined and the capacity of the mined material itself will determine the pillar size.

If one pillar fails and surrounding pillars are unable to support the area previously supported by the failed pillar they may in turn fail. This could lead to the collapse of the whole mine. To prevent this the mine is divided up into areas or panels. Pillars known as barrier pillars separate the panels. The barrier pillars are significantly larger than the "panel" pillars and are sized to allow them to support a significant part of the panel and prevent progressive collapse of the mine in the event of failure of the panel pillars.

Retreat Mining

Retreat mining is often the final stage of room and pillar mining. Once a deposit has been exhausted using this method, the pillars that were left behind initially are removed, or "pulled", retreating back towards the mine's entrance. After the pillars are removed, the roof (or back) is allowed to collapse behind the mining area. Pillar removal must occur in a very precise order to reduce the risks to workers, owing to the high stresses placed on the remaining pillars by the abutment stresses of the caving ground.

Retreat mining is a particularly dangerous form of mining: according to the Mine Safety and Health Administration (MSHA), pillar recovery mining has been historically responsible for 25% of American coal mining deaths caused by failures of the roof or walls, even though it represents only 10% of the coal mining industry.

Longwall Mining

Longwall mining

Longwall mining is a form of underground coal mining where a long wall of coal is mined in a single slice (typically 0.6 – 1.0 m thick). The longwall *panel* (the block of coal that is being mined) is typically 3 – 4 km long and 250 – 400 m wide.

History

Plan of longwall mine before conveyors, hoist is at the center of the central pillar.

The basic idea of longwall mining was developed in England in the late 17th century. Miners would undercut the coal along the width of the coal face, removing coal as it fell, and using wooden props to control the fall of the roof behind the face. This was known as the *Shropshire method* of mining. While the technology has changed considerably, the basic idea remains the same, to remove essentially all of the coal from a broad coal face and allow the roof and overlying rock to collapse into the void behind, while maintaining a safe working space along the face for the miners.

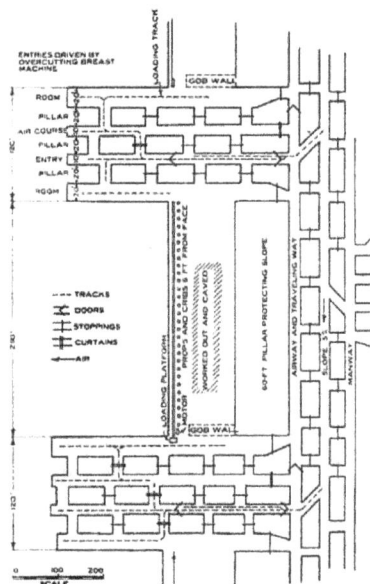

Oklahoma advancing longwall mine c. 1917; arrows show airflow

West Virginia retreating longwall mine c. 1917

Starting around 1900, mechanization was applied to this method. By 1940, some referred to long-wall mining as "the conveyor method" of mining, after the most prominent piece of machinery involved. Unlike earlier longwall mining, the use of a conveyor belt parallel to the coal face forced the face to be developed along a straight line. The only other machinery used was an electric cutter to undercut the coal face and electric drills for blasting to drop the face. Once dropped, manual labor was used to load coal onto the conveyor parallel to the face and to place wooden roof props to control the fall of the roof.

Such low-technology longwall mines continued in operation into the 1970s. The best known example was the New Gladstone Mine near Centerville, Iowa "one of the last advancing longwall mines in the United States". This longwall mine did not even use a conveyor belt, but relied on ponies to haul coal tubs from the face to the slope where a hoist hauled the tubs to the surface.

Longwall mining has been extensively used as the final stage in mining old room and pillar mines. In this context, longwall mining can be classified as a form of retreat mining.

Layout

Gate roads are driven to the back of each panel before longwall mining begins. The gate road along one side of the block is called the *maingate* or *headgate*; the road on the other side is called the *tailgate*. Where the thickness of the coal allows, these gate roads have been previously developed by continuous miner units, as the longwall itself is not capable of the initial development. The layout of Longwall could be either 'advancing' type or of 'retreat' type. In the advancing type, the gate roads are formed as the coal face advances. In thinner seams the advancing longwall mining method may be used. In the retreat type, the panel is formed by driving maingate, tail gate and a face connecting the both. Only the maingate road is formed in advance of the face. The tailgate road is formed behind the coal face by removing the stone above coal height to form a roadway that is high enough to travel in. The end of the block that includes the longwall equipment is called the face. The other end of the block is usually one of the main travel roads of the mine. The cavity behind the longwall is called the *goaf*, *goff* or *gob*.

Ventilation

Typically, intake (fresh) air travels up the main gate, across the face, and then down the tail gate, known as 'U' type ventilation. Once past the face the air is no longer fresh air, but return air carrying away coal dust and mine gases such as methane, carbon dioxide, depending on the geology of the coal. Return air is extracted by ventilation fans mounted on the surface. Other ventilation methods can be used where intake air also passes the main gate and into a bleeder or back return road reducing gas emissions from the goaf onto the face, or intake air travels up the tail gate and across the face in the same direction as the face chain in a homotropal system.

To avoid spontaneous combustion of coal in the goaf area, gases may be allowed to build up behind seals so as to exclude oxygen from the sealed goaf area. Where a goaf may contain an explosive mixture of methane and oxygen, nitrogen injection/inertisation may be used to exclude oxygen or push the explosive mixture deep into the goaf where there are no probable ignition sources. Seals are required to be monitored each shift by a certified mine supervisor for damage and leaks of harmful gases.

Equipment

Hydraulic chocks

A number of hydraulic jacks, called *powered roof supports*, *chocks* or *shields*, which are typically 1.75 m wide and placed in a long line, side by side for up to 400 m in length in order to support the roof of the coalface. An individual chock can weigh 30–40 tonnes, extend to a maximum cutting height of up to 6 m and have yield rating of 1000–1250 tonnes each, and hydraulically advance itself 1 m at a time.

Hydraulic chocks, conveyor and shearer

The coal is cut from the coalface by a machine called the *shearer* (*power loader*). This machine can weigh 75–120 tonnes typically and comprises a main body, housing the electrical functions, the tractive motive units to move the shearer along the coalface and pumping units (to power both hydraulic and water functions). At either end of the main body are fitted the ranging arms which can be ranged vertically up down by means of hydraulic rams, and onto which are mounted the shearer cutting drums which are fitted with 40–60 cutting picks. Within the ranging arms are housed very powerful electric motors (typically up to 850 kW) which transfer their power through a series of lay gears within the body and through the arms to the drum mounting locations at the extreme ends of the ranging arms where the cutting drums are. The cutting drums are rotated at a speed of 20–50 revs/min to cut the mineral from coal seam.

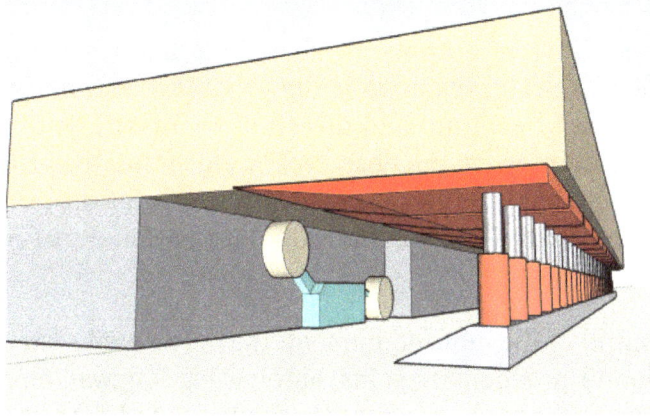

Chocks providing support to allow shearer to work

The shearer is carried along the length of the face on the *armoured face conveyor* (AFC); using a chain-less haulage system, which resembles a ruggedised rack and pinion system specially developed for mining. Prior to the chainless haulage systems, haulage systems with Chain were popular, where a heavy duty chain was run along the length of the coal face for the shearer to pull itself along. The shearer moves at a speed of 10–30 m/min depending on cutting conditions.

The AFC is placed in front of the powered roof supports, and the shearing action of the rotating drums cutting into the coal seam disintegrates the coal, this being loaded onto the AFC. The coal is removed from the coal face by a scraper chain conveyor to the main gate. Here it is loaded onto a network of conveyor belts for transport to the surface. At the main gate the coal is usually reduced in size in a crusher, and loaded onto the first conveyor belt by the *beam stage loader* (BSL).

As the shearer removes the coal, the AFC is snaked over behind the shearer and the powered roof supports move forward into the newly created cavity. As mining progresses and the entire long-wall progresses through the seam, the goaf increases. This goaf collapses under the weight of the overlying strata. The strata approximately 2.5 times the thickness of the coal seam removed collapses and the beds above settle onto the collapsed goaf. This collapsing can lower surface height, causing problems such as changing the course of rivers and severely damaging building foundations.

Comparison with Room and Pillar Method

Longwall and room and pillar methods of mining can both be used for mining suitable underground coal seams. Longwall has better resource recovery (about 80% compared with about 60% for room and pillar method, fewer roof support consumables are needed, higher volume coal clearance systems, minimal manual handling and safety of the miners is enhanced by the fact that they are always under the hydraulic roof supports when they are extracting coal.

Longwall Mining Automation

Longwall mining has traditionally been a manual process in which alignment of the face equipment was done with string lines. Technologies have been developed which automates several aspects of the longwall mining operation, including a system that aligns the face of the retreating longwall panel perpendicularly to the gate-roads.

Briefly, Inertial navigation system outputs are used in a dead-reckoning calculation to estimate the shearer positions. (Dead reckoning is the process of estimating the current position based on previous estimates of position and direction of travel.) Optimal Kalman filters and smoothers can be applied to improve the dead reckoning estimates prior to repositioning the longwall equipment at the completion of each shear . Expectation-maximization algorithms can be used to estimate the unknown filter and smoother parameters for tracking the longwall shearer positions.

Compared to manual control of the mine equipment, the automated system yields improved production rates. In addition to productivity gains, automating longwall equipment leads to safety benefits. The coalface is a hazardous area because methane and carbon monoxide are present, while the area is hot and humid since water is sprayed over the face to minimize the likelihood of sparks occurring when the shearer picks strike rock. By automating manual processes, face workers can be removed from these hazardous areas.

Subsidence

There have been cases of surface subsidence altering the landscape above the mines. At Newstan Colliery in New South Wales, Australia "the surface has dropped by as much as five metres in places" above a multi level mine. In some cases the subsidence causes damage to natural features such as drainage to water courses or man-made structures such as roads and buildings. "Douglas Park Drive was closed for four weeks because longwall panels ... destabilised the road. In 2000, the State Government stopped mining when it came within 600 metres from the twin bridges. A year later there were reports of 40-centimetre gaps appearing in the road, and the bridge had to be jacked sideways to realign it." p. 2

A 2005 geotechnical report commissioned by the NSW RTA warns that "subsidence could happen suddenly and occur over many years".

However, there are several mines, which were successfully mined with little to no measurable surface subsidence including mines under lakes, oceans, important water catchments and environmentally sensitive areas. Subsidence is minimised by increasing the block's adjacent chain pillar widths, decreasing extracted block widths and heights, and by giving consideration to the depth of cover as well as competency and thickness of overlying strata.

Shaft Mining

Abandoned mine shafts in Marl, Germany.

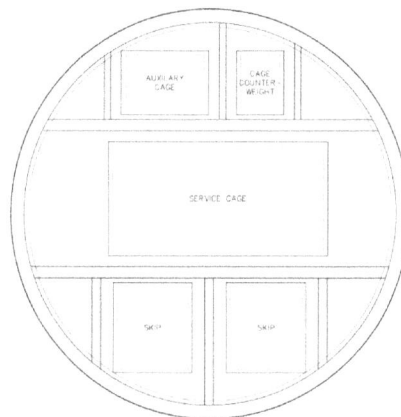

A plan-view schematic of a mine shaft showing cage and skip compartments. Services may be housed in either of the four open compartments.

Shaft mining or shaft sinking is excavating a vertical or near-vertical tunnel from the top down, where there is initially no access to the bottom.

Shallow shafts, typically sunk for civil engineering projects differ greatly in execution method from deep shafts, typically sunk for mining projects. When the top of the excavation is the ground surface, it is referred to as a shaft; when the top of the excavation is underground, it is called a winze or a sub-shaft. Small shafts may be excavated upwards from within an existing mine as long as there is access at the bottom, in which case they are called Raises. A shaft may be either vertical or inclined (between 45 and 90 degrees to the horizontal), although most modern mine shafts are vertical. If access exists at the bottom of the proposed shaft and ground conditions allow then raise boring may be used to excavate the shaft from the bottom up, such shafts are called *borehole shafts*. Shaft Sinking is one of the most difficult of all development methods: restricted space, gravity, groundwater and specialized procedures make the task quite formidable.

Historically mine shaft sinking has been among the most dangerous of all mining occupations and the preserve of mining contractors. Today shaft sinking contractors are concentrated in Canada and South Africa.

Parts of a Mine Shaft

The most visible feature of a mine shaft is the headframe (or winding tower, poppet head or pit head) which stands above the shaft. Depending on the type of hoist used the top of the headframe will either house a hoist motor or a sheave wheel (with the hoist motor mounted on the ground). The headframe will also contain bins for storing ore being transferred to the processing facility.

At ground level beneath and around the headframe is the Shaft Collar (also called the Bank or Deck), which provides the foundation necessary to support the weight of the headframe and provides a means for men, materials and services to enter and exit the shaft. Collars are usually massive reinforced concrete structures with more than one level. If the shaft is used for mine ventilation, a plenum space or casing is incorporated into the collar to ensure the proper flow of air into and out of the mine.

Beneath the Collar the part of the shaft which continues into the ground is called the Shaft Barrel.

At locations where the Shaft Barrel meets horizontal workings there is a Shaft Station which allows men, materials and services to enter and exit the shaft. From the station tunnels (drifts, galleries or levels) extend towards the ore body, sometimes for many kilometers. The lowest Shaft Station is most often the point where rock leaves the mine levels and is transferred to the shaft, if so a Loading Pocket is excavated on one side of the shaft at this location to allow transfer facilities to be built.

Beneath the lowest Shaft Station the shaft continues on for some distance, this area is referred to as the Shaft Bottom. A tunnel called a Ramp typically connects the bottom of the shaft with the rest of the mine, this Ramp often contains the mine's water handling facility, called the Sump, as water will naturally flow to the lowest point in the mine.

Shafts may be sunk by conventional drill and blast or mechanised means. The industry is gradually attempting to shift further towards shaft boring but a reliable method to do so has yet to be developed.

Shaft Lining

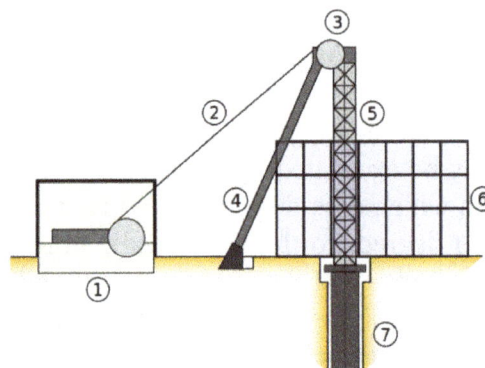

Schematic of headframe

1. hoist
2. cable
3. wheel
4. sheer
5. false edge
6. hoistroom
7. mineshaft

Pulley wheel of 1 Maja Coal Mine in Wodzisław Śląski

The shaft lining performs several functions; it is first and foremost a safety feature preventing loose or unstable rock from falling into the shaft, then a place for shaft sets to bolt into and lastly a smooth surface to minimise resistance to airflow for ventilation.

In North and South America, smaller shafts are designed to be rectangular with timber supports. Larger shafts are round and are concrete lined.

Final choice of shaft lining is dependent on the geology of the rock which the shaft passes through, some shafts have several liners sections as required Where shafts are sunk in very competent rock there may be no requirement for lining at all, or just the installation of welded mesh and rock bolts. The material of choice for shaft lining is mass concrete which is poured behind Shaft Forms in Lifts of 6m as the shaft advances (gets deeper).

Shotcrete, fibrecrete, brick, cast iron tubing, precast concrete segments have all been used at one time or another. Additionally, the use of materials like Bitumen and even squash balls have been required by specific circumstances. In extreme circumstances, particularly when sinking through Halite, composite liners consisting of two or more materials may be required.

The shaft liner does not. reach right to the bottom of the shaft during sinking, but lags behind by a fixed distance. This distance is determined by the methodology of excavation and the design thickness of the permanent liner. To ensure the safety of persons working on the shaft bottom temporary ground support is installed, usually consisting of welded mesh and rock bolts. The installation of the temporary ground support (called *bolting*) is among the most physically challenging parts of the shaft sinking cycle as bolts must be installed using pneumatic powered rock drills.

For this reason, and to minimise the number of persons on the shaft bottom a number of projects have successfully switched to shotcrete for this temporary lining. Research and development in this area is focusing on the robotic application of shotcrete and the commercialisation of thin sprayed polymer liners.

Shaft Compartments

Typical mine cage, located in Harzbergbau, Germany

Where the shaft is to be used for hoisting it is frequently split into multiple compartments by Shaft Sets, these may be made of either timber or steel. Vertical members in a shaft set are called Guides, horizontal members are called Buntons. For steel shaft guides, the main two options are hollow structural sections and top hat sections. Top hat sections offer a number of advantages over hollow structural sections including simpler installation, improved corrosion resistance and increased stiffness. Mine conveyances run on the guides in a similar way to how a steel roller coaster runs on its rails, both having wheels which keep them securely in place.

Some shafts do not use guide beams but instead utilize steel wire rope (called Guide ropes) kept in tension by massive weights at shaft bottom called cheese weights (because of their resemblance to a truckle or wheel of cheese) as these are easier to maintain and replace.

The largest compartment is typically used for the mine cage, a conveyance used for moving workers and supplies below the surface, which is suspended from the hoist on steel wire rope. It functions in a similar manner to an elevator. Cages may be single-, double-, or rarely triple-deck, always having multiple redundant safety systems in case of unexpected failure.

The second compartment is used for one or more skips, used to hoist ore to the surface. Smaller mining operations use a skip mounted underneath the cage, rather than a separate device, while some large mines have separate shafts for the cage and skips. The third compartment is used for an emergency exit; it may house an auxiliary cage or a system of ladders. An additional compartment houses mine services such as high voltage cables and pipes for transfer of water, compressed air or diesel fuel.

A second reason to divide the shaft is for ventilation. One or more of the compartments discussed above may be used for air intake, while others may be used for exhaust. Where this is the case a steel wall called a Brattice is installed between the two compartments to separate the air flow. At many mines there are one or more complete additional separate "Auxiliary" shafts with separate head gear and cages. It is safer to have an alternate route to exit the mine as any problem in one shaft may affect all the compartments.

Drilling and Blasting

Rock blasting in Finland

Drilling and Blasting is the controlled use of explosives and other methods such as gas pressure blasting pyrotechnics, to break rock for excavation. It is practiced most often in mining, quarrying and civil engineering such as dam or road construction. The result of rock blasting is often known as a rock cut.

Drilling and Blasting currently utilizes many different varieties of explosives with different compositions and performance properties. Higher velocity explosives are used for relatively hard rock in order to shatter and break the rock, while low velocity explosives are used in soft rocks to generate more gas pressure and a greater heaving effect. For instance, an early 20th-century blasting manual compared the effects of black powder to that of a wedge, and dynamite to that of a hammer. The most commonly used explosives in mining today are ANFO based blends due to lower cost than dynamite.

Before the advent of tunnel boring machines, drilling and blasting was the only economical way of excavating long tunnels through hard rock, where digging is not possible. Even today, the method is still used in the construction of tunnels, such as in the construction of the Lötschberg Base Tunnel. The decision whether to construct a tunnel using a TBM or using a drill and blast method includes a number of factors. Tunnel length is a key issue that needs to be addressed because large TBMs for a rock tunnel have a high capital cost, but because they are usually quicker than a drill and blast tunnel the price per metre of tunnel is lower. This means that shorter tunnels tend to be less economical to construct with a TBM and are therefore usually constructed by drill and blast. Managing ground conditions can also have a significant effect on the choice with different methods suited to different hazards in the ground.

History

The use of explosives in mining goes back to the year 1627, when gunpowder was first used in place of mechanical tools in the Hungarian (now Slovakian) town of Banská Štiavnica. The innovation spread quickly throughout Europe and the Americas.

While drilling and blasting saw limited use in pre-industrial times using gunpowder (such as with the Blue Ridge Tunnel in the United States, built in the 1850s), it was not until more powerful (and safer) explosives, such as dynamite (patented 1867), as well as powered drills were developed, that its potential was fully realised.

Drilling and blasting was successfully used to construct tunnels throughout the world, notably the Fréjus Rail Tunnel, the Gotthard Rail Tunnel, the Simplon Tunnel, the Jungfraubahn and even the longest road tunnel in the world, Lærdalstunnelen, are constructed using this method.

In 1990, 2.1 million tonnes (2.32 million short tons) of commercial explosives were consumed in the United States, representing an estimated expenditure of 3.5 to 4 billion 1993 dollars on blasting. Australia had the highest explosives consumption that year at 500 million tonnes (551 million short tons), with Scandinavian countries another leader in rock blasting (Persson et al. 1994:1).

Procedure

A drill jumbo during the construction of Citybanan under Stockholm, used for drilling holes for explosives

As the name suggests, drilling and blasting works as follows:

- A number of holes are drilled into the rock, which are then filled with explosives.

- Detonating the explosive causes the rock to collapse.

- Rubble is removed and the new tunnel surface is reinforced.

- Repeating these steps until desired excavation is complete.

The positions and depths of the holes (and the amount of explosive each hole receives) are determined by a carefully constructed pattern, which, together with the correct timing of the individual explosions, will guarantee that the tunnel will have an approximately circular cross-section.

During operation, blasting mats may be used to contain the blast, suppress dust and noise, for fly rock prevention and sometimes to direct the blast.

Rock Support

As a tunnel or excavation progresses the roof and side walls of need to be supported to stop the rock falling into the excavation. The philosophy and methods for rock support vary widely but typical rock support systems can include:

- Rock bolts or rock dowels

- Shotcrete

- Ribs or mining arches and lagging

- Cable bolts

- In-situ concrete

Typically a rock support system would include a number of these support methods, each intended to undertake a specific role in the rock support such as the combination of rock bolting and shotcrete.

References

- Anderson, Wayne I. (1998). Iowa's geological past: three billion years of earth history. Iowa City 52242: University of Iowa Press. p. 258. ISBN 0-87745-639-9. Retrieved 2010-06-05.

- Brick, Greg A. (2004). Iowa Underground: a guide to the state's subterranean treasures. Black Earth, Wis.: Trails Books. pp. 119–120. ISBN 978-1-931599-39-9. OL 3314195M.

- De la Vergne, Jack (August 2003). Hard Rock Miner's Handbook, Edition 3. Tempe/North Bay: McIntosh Engineering. p. 92. ISBN 0-9687006-1-6.

- DellaMea, Chris. "Upper Youghiogheny Coalfield." Coalfields of the Appalachian Mountains. November 1, 2006. Accessed December 28, 2015.

- DellaMea, Chris. "Upper Youghiogheny Coalfield." Coalfields of the Appalachian Mountains. November 1, 2006. Accessed December 28, 2015.

- DellaMea, Chris. "Eastern Kentucky Coalfields." Coalfields of the Appalachian Mountains. November 1, 2006. Accessed December 28, 2015.

- Washlaski, Raymond, and Ryan Washlaski. "Virtual Museum of Coal Mining in Western Pennsylvania." September 18, 2010. Accessed December 28, 2015.

- DellaMea, Chris. "Eastern Tennessee Coalfields." Coalfields of the Appalachian Mountains. November 1, 2006. Accessed December 28, 2015.

- Silva, Michael. "Placer Gold Recovery Methods (SP87)." California Department of Conservation Division of Mines and Geology. 1986. Accessed December 28, 2015.

- Sames, Gary, and Noel Moebs. "Hillseam Geology and Roof Instability near Outcrop in Eastern Kentucky Drift Mines." U.S. Bureau of Mines. 1989. Accessed December 28, 2015.

Minerals Extracted Through Mining

The major components discussed in this chapter are coal, oil shale and clay. This chapter helps the reader to develop a comprehensive understanding on the minerals extracted through mining. The aspects elucidated are of vital importance and provide a better understanding on mining.

Coal

Coal (Old English *col*; meaning "mineral of fossilized carbon" since the thirteenth century) is a combustible black or brownish-black sedimentary rock usually occurring in rock strata in layers or veins called coal beds or coal seams. The harder forms, such as anthracite coal, can be regarded as metamorphic rock because of later exposure to elevated temperature and pressure. Coal is composed primarily of carbon along with variable quantities of other elements, chiefly hydrogen, sulfur, oxygen, and nitrogen. A fossil fuel, coal forms when dead plant matter is converted into peat, which in turn is converted into lignite, then sub-bituminous coal, after that bituminous coal, and lastly anthracite. This involves biological and geological processes that take place over a long period.

Bituminous coal

Throughout history, coal has been used as an energy resource, primarily burned for the production of electricity and/or heat, and is also used for industrial purposes, such as refining metals. Coal is the largest source of energy for the generation of electricity worldwide, as well as one of the largest worldwide anthropogenic sources of carbon dioxide releases. The extraction of coal, its use in energy production and its byproducts are all associated with environmental and health effects including Climate change.

Coal is extracted from the ground by coal mining. Since 1983, the world's top coal producer has been China. In 2011 China produced 3,520 million tonnes of coal – 49.5% of 7,695 million tonnes world coal production. In 2011 other large producers were United States (993 million tonnes), India (589), European Union (576) and Australia (416). In 2010 the largest exporters were Australia with 328 million tonnes (27.1% of world coal export) and Indonesia with 316 million tonnes (26.1%), while the largest importers were Japan with 207 million tonnes (17.5% of world coal import), China with 195 million tonnes (16.6%) and South Korea with 126 million tonnes (10.7%).

Etymology

The word originally took the form *col* in Old English, from Proto-Germanic **kula(n)*, which in turn is hypothesized to come from the Proto-Indo-European root **g(e)u-lo-* "live coal". Germanic cognates include the Old Frisian *kole*, Middle Dutch *cole*, Dutch *kool*, Old High German *chol*, German *Kohle* and Old Norse *kol*, and the Irish word *gual* is also a cognate via the Indo-European root. In Old Turkic languages, *kül* is "ash(es), cinders", *öčür* is "quench". The compound "charcoal" in Turkic is *öčür(ülmüş) kül*, literally "quenched ashes, cinders, coals" with elided anlaut *ö*- and inflection affixes *-ülmüş*.

The word took on the meaning "mineral consisting of fossilized carbon" in the thirteenth century. Often the terms "coal plant" or "coal power plant" are used, referring to the presumed origin of coal being old fossilized plants.

Formation

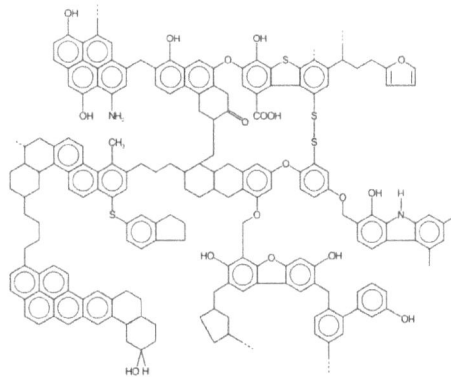

Example chemical structure of coal

At various times in the geologic past, the Earth had dense forests in low-lying wetland areas. Due to natural processes such as flooding, these forests were buried underneath soil. As more and more soil deposited over them, they were compressed. The temperature also rose as they sank deeper and deeper. As the process continued the plant matter was protected from biodegradation and oxidation, usually by mud or acidic water. This trapped the carbon in immense peat bogs that were eventually covered and deeply buried by sediments. Under high pressure and high temperature, dead vegetation was slowly converted to coal. As coal contains mainly carbon, the conversion of dead vegetation into coal is called carbonization.

The wide, shallow seas of the Carboniferous Period provided ideal conditions for coal formation, although coal is known from most geological periods. The exception is the coal gap in the Permian–Triassic extinction event, where coal is rare. Coal is known from Precambrian strata, which predate land plants — this coal is presumed to have originated from residues of algae.

Ranks

Coastal exposure of the Point Aconi Seam (Nova Scotia)

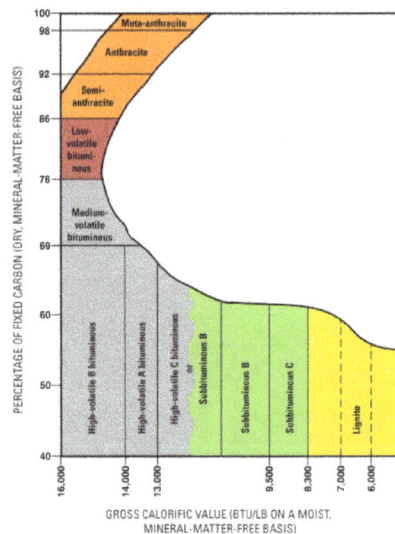

Coal ranking system used in the United States (US Geological Survey)

As geological processes apply pressure to dead biotic material over time, under suitable conditions, its metamorphic grade increases successively into:

- Peat, considered to be a precursor of coal, has industrial importance as a fuel in some regions, for example, Ireland and Finland. In its dehydrated form, peat is a highly effective absorbent for fuel and oil spills on land and water. It is also used as a conditioner for soil to make it more able to retain and slowly release water.

- Lignite, or brown coal, is the lowest rank of coal and used almost exclusively as fuel for electric power generation. Jet, a compact form of lignite, is sometimes polished and has been used as an ornamental stone since the Upper Palaeolithic.

- Sub-bituminous coal, whose properties range from those of lignite to those of bituminous coal, is used primarily as fuel for steam-electric power generation and is an important source of light aromatic hydrocarbons for the chemical synthesis industry.

- Bituminous coal is a dense sedimentary rock, usually black, but sometimes dark brown, often with well-defined bands of bright and dull material; it is used primarily as fuel in steam-electric power generation, with substantial quantities used for heat and power applications in manufacturing and to make coke.

- "Steam coal" is a grade between bituminous coal and anthracite, once widely used as a fuel for steam locomotives. In this specialized use, it is sometimes known as "sea coal" in the US. Small steam coal (dry small steam nuts or DSSN) was used as a fuel for domestic water heating.

- Anthracite, the highest rank of coal, is a harder, glossy black coal used primarily for residential and commercial space heating. It may be divided further into metamorphically altered bituminous coal and "petrified oil", as from the deposits in Pennsylvania.

- Graphite is one of the more difficult coals to ignite and is not commonly used as fuel — it is mostly used in pencils, and when powdered, as a lubricant.

The classification of coal is generally based on the content of volatiles. However, the exact classification varies between countries. According to the German classification, coal is classified as follows:

German Classification	English Designation	Volatiles %	C Carbon %	H Hydrogen %	O Oxygen %	S Sulfur %	Heat content kJ/kg
Braunkohle	Lignite (brown coal)	45–65	60–75	6.0–5.8	34-17	0.5-3	<28,470
Flammkohle	Flame coal	40-45	75-82	6.0-5.8	>9.8	~1	<32,870
Gasflammkohle	Gas flame coal	35-40	82-85	5.8-5.6	9.8-7.3	~1	<33,910
Gaskohle	Gas coal	28-35	85-87.5	5.6-5.0	7.3-4.5	~1	<34,960
Fettkohle	Fat coal	19-28	87.5-89.5	5.0-4.5	4.5-3.2	~1	<35,380
Esskohle	Forge coal	14-19	89.5-90.5	4.5-4.0	3.2-2.8	~1	<35,380
Magerkohle	Nonbaking coal	10-14	90.5-91.5	4.0-3.75	2.8-3.5	~1	35,380
Anthrazit	Anthracite	7-12	>91.5	<3.75	<2.5	~1	<35,300
Note, the percentages are percent by mass of the indicated elements							

The middle six grades in the table represent a progressive transition from the English-language sub-bituminous to bituminous coal, while the last class is an approximate equivalent to anthracite, but more inclusive (US anthracite has < 6% volatiles).

Cannel coal (sometimes called "candle coal") is a variety of fine-grained, high-rank coal with significant hydrogen content. It consists primarily of "exinite" macerals, now termed "liptinite".

Hilt's Law

Hilt's law is a geological term that states that, in a small area, the deeper the coal, the higher its rank (grade). The law holds true if the thermal gradient is entirely vertical, but metamorphism may cause lateral changes of rank, irrespective of depth.

Early Uses as Fuel

Chinese coal miners in an illustration of the *Tiangong Kaiwu* encyclopedia, published in 1637

The earliest recognized use is from the Shenyang area of China 4000 BC where Neolithic inhabitants had begun carving ornaments from black lignite. Coal from the Fushun mine in northeastern China was used to smelt copper as early as 1000 BCE. Marco Polo, the Italian who traveled to China in the 13th century, described coal as "black stones ... which burn like logs", and said coal was so plentiful, people could take three hot baths a week. In Europe, the earliest reference to the use of coal as fuel is from the geological treatise *On stones* (Lap. 16) by the Greek scientist Theophrastus (*circa* 371–287 BC):

Among the materials that are dug because they are useful, those known as *anthrakes* [coals] are made of earth, and, once set on fire, they burn like charcoal. They are found in Liguria ... and in Elis as one approaches Olympia by the mountain road; and they are used by those who work in metals.

—Theophrastus, On Stones (16) translation

Outcrop coal was used in Britain during the Bronze Age (3000–2000 BC), where it has been detected as forming part of the composition of funeral pyres. In Roman Britain, with the exception of two modern fields, "the Romans were exploiting coals in all the major coalfields in England and Wales by the end of the second century AD". Evidence of trade in coal (dated to about AD 200) has been found at the Roman settlement at Heronbridge, near Chester, and in the Fenlands of East Anglia, where coal from the Midlands was transported via the Car Dyke for use in drying grain. Coal cinders have been found in the hearths of villas and Roman forts, particularly in Northumberland, dated to around AD 400. In the west of England, contemporary writers described the wonder of a permanent brazier of coal on the altar of Minerva at Aquae Sulis (modern day Bath), although in fact easily accessible surface coal from what became the Somerset coalfield was in common use in quite lowly dwellings locally. Evidence of coal's use for iron-working in the city during the Roman period has been found. In Eschweiler, Rhineland, deposits of bituminous coal were used by the Romans for the smelting of iron ore.

Coal miner in Britain, 1942

No evidence exists of the product being of great importance in Britain before the High Middle Ages, after about AD 1000. Mineral coal came to be referred to as "seacoal" in the 13th century; the wharf where the material arrived in London was known as Seacoal Lane, so identified in a charter of King Henry III granted in 1253. Initially, the name was given because much coal was found on the shore, having fallen from the exposed coal seams on cliffs above or washed out of underwater coal outcrops, but by the time of Henry VIII, it was understood to derive from the way it was carried to London by sea. In 1257–59, coal from Newcastle upon Tyne was shipped to London for the smiths and lime-burners building Westminster Abbey. Seacoal Lane and Newcastle Lane, where coal was unloaded at wharves along the River Fleet, are still in existence.

These easily accessible sources had largely become exhausted (or could not meet the growing demand) by the 13th century, when underground extraction by shaft mining or adits was developed. The alternative name was "pitcoal", because it came from mines. It was, however, the development of the Industrial Revolution that led to the large-scale use of coal, as the steam engine took over from the water wheel. In 1700, five-sixths of the world's coal was mined in Britain. Britain would have run out of suitable sites for watermills by the 1830s if coal had not been available as a source of energy. In 1947, there were some 750,000 miners in Britain, but by 2004, this had shrunk to some 5,000 miners working in around 20 collieries.

Uses Today

Castle Gate Power Plant near Helper, Utah, USA

Coal rail cars

Coal as Fuel

Coal is primarily used as a solid fuel to produce electricity and heat through combustion. World coal consumption was about 7.25 billion tonnes in 2010 (7.99 billion short tons) and is expected to increase 48% to 9.05 billion tonnes (9.98 billion short tons) by 2030. China produced 3.47 billion tonnes (3.83 billion short tons) in 2011. India produced about 578 million tonnes (637.1 million short tons) in 2011. 68.7% of China's electricity comes from coal. The USA consumed about 13% of the world total in 2010, i.e. 951 million tonnes (1.05 billion short tons), using 93% of it for generation of electricity. 46% of total power generated in the USA was done using coal. The United States Energy Information Administration estimates coal reserves at 948×10^9 short tons (860 Gt). One estimate for resources is 18,000 Gt.

When coal is used for electricity generation, it is usually pulverized and then combusted (burned) in a furnace with a boiler. The furnace heat converts boiler water to steam, which is then used to spin turbines which turn generators and create electricity. The thermodynamic efficiency of this process has been improved over time; some older coal-fired power stations have thermal efficiencies in the vicinity of 25% whereas the newest supercritical and "ultra-supercritical" steam cycle turbines, operating at temperatures over 600 °C and pressures over 27 MPa (over 3900 psi), can practically achieve thermal efficiencies in excess of 45% (LHV basis) using anthracite fuel, or around 43% (LHV basis) even when using lower-grade lignite fuel. Further thermal efficiency improvements are also achievable by improved pre-drying (especially relevant with high-moisture fuel such as lignite or biomass) and cooling technologies.

An alternative approach of using coal for electricity generation with improved efficiency is the integrated gasification combined cycle (IGCC) power plant. Instead of pulverizing the coal and burning it directly as fuel in the steam-generating boiler, the coal can be first gasified to create syngas, which is burned in a gas turbine to produce electricity (just like natural gas is burned in a turbine). Hot exhaust gases from the turbine are used to raise steam in a heat recovery steam generator which powers a supplemental steam turbine. Thermal efficiencies of current IGCC power plants range from 39-42% (HHV basis) or ~42-45% (LHV basis) for bituminous coal and assuming utilization of mainstream gasification technologies (Shell, GE Gasifier, CB&I). IGCC power plants outperform conventional pulverized coal-fueled plants in terms of pollutant emissions, and allow for relatively easy carbon capture.

At least 40% of the world's electricity comes from coal, and in 2012, about one-third of the United States' electricity came from coal, down from approximately 49% in 2008. As of 2012 in the United States, use of coal to generate electricity was declining, as plentiful supplies of natural gas obtained by hydraulic fracturing of tight shale formations became available at low prices.

In Denmark, a net electric efficiency of > 47% has been obtained at the coal-fired Nordjyllandsværket CHP Plant and an overall plant efficiency of up to 91% with cogeneration of electricity and district heating. The multifuel-fired Avedøreværket CHP Plant just outside Copenhagen can achieve a net electric efficiency as high as 49%. The overall plant efficiency with cogeneration of electricity and district heating can reach as much as 94%.

An alternative form of coal combustion is as coal-water slurry fuel (CWS), which was developed in the Soviet Union. CWS significantly reduces emissions, improving the heating value of coal. Other ways to use coal are combined heat and power cogeneration and an MHD topping cycle.

The total known deposits recoverable by current technologies, including highly polluting, low-energy content types of coal (i.e., lignite, bituminous), is sufficient for many years. However, consumption is increasing and maximal production could be reached within decades. On the other hand, much may have to be left in the ground to avoid climate change.

Coking Coal and Use of Coke

Coke oven at a smokeless fuel plant in Wales, United Kingdom

Coke is a solid carbonaceous residue derived from low-ash, low-sulfur bituminous coal (metallurgical coal), from which the volatile constituents are driven off by baking in an oven without oxygen at temperatures as high as 1,000 °C (1,832 °F), so the fixed carbon and residual ash are fused together. Metallurgical coke is used as a fuel and as a reducing agent in smelting iron ore in a blast furnace. The result is pig iron, and is too rich in dissolved carbon, so it must be treated further to make steel. The coking coal should be low in sulfur and phosphorus, so they do not migrate to the metal. Based on the ash percentage, the coking coal can be divided into various grades. These grades are:

- Steel Grade I (Not exceeding 15%)

- Steel Grade II (Exceeding 15% but not exceeding 18%)

- Washery Grade I (Exceeding 18% but not exceeding 21%)

- Washery Grade II (Exceeding 21% but not exceeding 24%)

- Washery Grade III (Exceeding 24% but not exceeding 28%)

- Washery Grade IV (Exceeding 28% but not exceeding 35%)

The coke must be strong enough to resist the weight of overburden in the blast furnace, which is why coking coal is so important in making steel using the conventional route. However, the alternative route is direct reduced iron, where any carbonaceous fuel can be used to make sponge or pelletised iron. Coke from coal is grey, hard, and porous and has a heating value of 24.8 million Btu/ton (29.6 MJ/kg). Some cokemaking processes produce valuable byproducts, including coal tar, ammonia, light oils, and coal gas.

Petroleum coke is the solid residue obtained in oil refining, which resembles coke, but contains too many impurities to be useful in metallurgical applications.

Gasification

Coal gasification can be used to produce syngas, a mixture of carbon monoxide (CO) and hydrogen (H_2) gas. Often syngas is used to fire gas turbines to produce electricity, but the versatility of syngas also allows it to be converted into transportation fuels, such as gasoline and diesel, through the Fischer-Tropsch process; alternatively, syngas can be converted into methanol, which can be blended into fuel directly or converted to gasoline via the methanol to gasoline process. Gasification combined with Fischer-Tropsch technology is currently used by the Sasol chemical company of South Africa to make motor vehicle fuels from coal and natural gas. Alternatively, the hydrogen obtained from gasification can be used for various purposes, such as powering a hydrogen economy, making ammonia, or upgrading fossil fuels.

During gasification, the coal is mixed with oxygen and steam while also being heated and pressurized. During the reaction, oxygen and water molecules oxidize the coal into carbon monoxide (CO), while also releasing hydrogen gas (H_2). This process has been conducted in both underground coal mines and in the production of town gas.

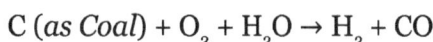

$$C \ (as \ Coal) + O_2 + H_2O \rightarrow H_2 + CO$$

If the refiner wants to produce gasoline, the syngas is collected at this state and routed into a Fischer-Tropsch reaction. If hydrogen is the desired end-product, however, the syngas is fed into the water gas shift reaction, where more hydrogen is liberated.

$$CO + H_2O \rightarrow CO_2 + H_2$$

In the past, coal was converted to make coal gas (town gas), which was piped to customers to burn for illumination, heating, and cooking.

Liquefaction

Coal can also be converted into synthetic fuels equivalent to gasoline or diesel by several different direct processes (which do not intrinsically require gasification or indirect conversion). In the direct liquefaction processes, the coal is either hydrogenated or carbonized. Hydrogenation processes are the Bergius process, the SRC-I and SRC-II (Solvent Refined Coal) processes, the NUS Corporation hydrogenation process and several other single-stage and two-stage processes. In the process of low-temperature carbonization, coal is coked at temperatures between 360 and 750 °C (680 and 1,380 °F). These temperatures optimize the production of coal tars richer in lighter hydrocarbons than normal coal tar. The coal tar is then further processed into fuels. An overview of coal liquefaction and its future potential is available.

Coal liquefaction methods involve carbon dioxide (CO_2) emissions in the conversion process. If coal liquefaction is done without employing either carbon capture and storage (CCS) technologies or biomass blending, the result is lifecycle greenhouse gas footprints that are generally greater than those released in the extraction and refinement of liquid fuel production from crude oil. If CCS technologies are employed, reductions of 5–12% can be achieved in Coal to Liquid (CTL) plants and up to a 75% reduction is achievable when co-gasifying coal with commercially demonstrated levels of biomass (30% biomass by weight) in coal/biomass-to-liquids plants. For future synthetic fuel projects, carbon dioxide sequestration is proposed to avoid releasing CO_2 into the atmosphere. Sequestration adds to the cost of production.

Refined Coal

Refined coal is the product of a coal-upgrading technology that removes moisture and certain pollutants from lower-rank coals such as sub-bituminous and lignite (brown) coals. It is one form of several precombustion treatments and processes for coal that alter coal's characteristics before it is burned. The goals of precombustion coal technologies are to increase efficiency and reduce emissions when the coal is burned. Depending on the situation, precombustion technology can be used in place of or as a supplement to postcombustion technologies to control emissions from coal-fueled boilers.

Industrial Processes

Finely ground bituminous coal, known in this application as sea coal, is a constituent of foundry sand. While the molten metal is in the mould, the coal burns slowly, releasing reducing gases at pressure, and so preventing the metal from penetrating the pores of the sand. It is also contained in 'mould wash', a paste or liquid with the same function applied to the mould before casting. Sea coal can be mixed with the clay lining (the "bod") used for the bottom of a cupola furnace. When heated, the coal decomposes and the bod becomes slightly friable, easing the process of breaking open holes for tapping the molten metal.

Production of Chemicals

Coal is an important feedstock in production of a wide range of chemical fertilizers and other chemical products. The main route to these products is coal gasification to produce syngas. Primary chemicals that are produced directly from the syngas include methanol, hydrogen and carbon monoxide, which are the chemical building blocks from which a whole spectrum of derivative chemicals are manufactured, including olefins, acetic acid, formaldehyde, ammonia, urea and others. The versatility of syngas as a precursor to primary chemicals and high-value derivative products provides the option of using relatively inexpensive coal to produce a wide range of valuable commodities.

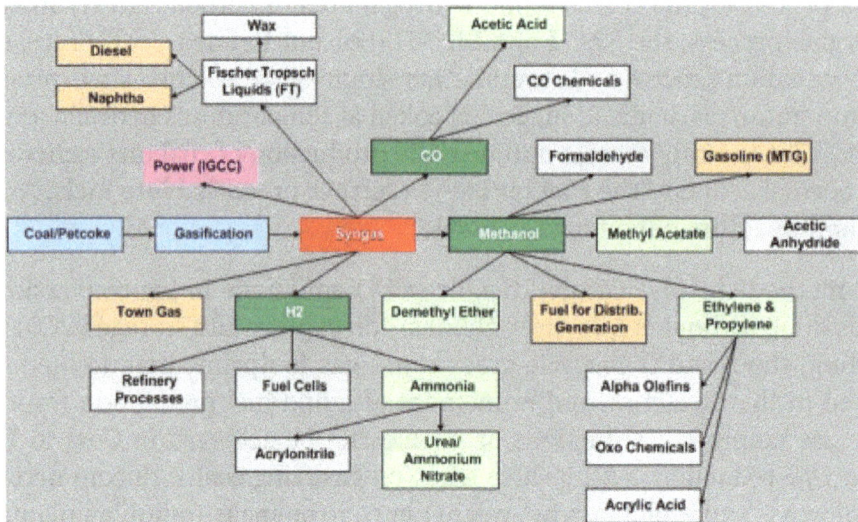

Production of Chemicals from Coal

Historically, production of chemicals from coal has been used since the 1950s and has become established in the market. According to the 2010 Worldwide Gasification Database, a survey of current and planned gasifiers, from 2004 to 2007 chemical production increased its gasification product share from 37% to 45%. From 2008 to 2010, 22% of new gasifier additions were to be for chemical production.

Because the slate of chemical products that can be made via coal gasification can in general also use feedstocks derived from natural gas and petroleum, the chemical industry tends to use whatever feedstocks are most cost-effective. Therefore, interest in using coal tends to increase for higher oil and natural gas prices and during periods of high global economic growth that may strain oil and gas production. Also, production of chemicals from coal is of much higher interest in countries like South Africa, China, India and the United States where there are abundant coal resources. The abundance of coal combined with lack of natural gas resources in China is strong inducement for the coal to chemicals industry pursued there. In the United States, the best example of the industry is Eastman Chemical Company which has been successfully operating a coal-to-chemicals plant at its Kingsport, Tennessee, site since 1983. Similarly, Sasol has built and operated coal-to-chemicals facilities in South Africa.

Coal to chemical processes do require substantial quantities of water. As of 2013 much of the coal to chemical production was in the People's Republic of China where environmental regulation and water management was weak.

Coal as a Traded Commodity

In North America, Central Appalachian coal futures contracts are currently traded on the New York Mercantile Exchange (trading symbol *QL*). The trading unit is 1,550 short tons (1,410 t) per contract, and is quoted in U.S. dollars and cents per ton. Since coal is the principal fuel for generating electricity in the United States, coal futures contracts provide coal producers and the electric power industry an important tool for hedging and risk management.

In addition to the NYMEX contract, the Intercontinental Exchange (ICE) has European (Rotterdam) and South African (Richards Bay) coal futures available for trading. The trading unit for these contracts is 5,000 tonnes (5,500 short tons), and are also quoted in U.S. dollars and cents per ton.

The price of coal increased from around $30.00 per short ton in 2000 to around $150.00 per short ton as of September 2008. As of October 2008, the price per short ton had declined to $111.50. Prices further declined to $71.25 as of October 2010. In early 2015, it was trading near $56/ton.

Environmental and Health Effects

Aerial photograph of Kingston Fossil Plant coal fly ash slurry spill site taken the day after the event

Health Effects

The use of coal as fuel causes adverse health impact and deaths.

The deadly London smog was caused primarily by the heavy use of coal. In the United States coal-fired power plants were estimated in 2004 to cause nearly 24,000 premature deaths every year, including 2,800 from lung cancer. Annual health costs in Europe from use of coal to generate electricity are €42.8 billion, or $55 billion. Yet the disease and mortality burden of coal use today falls most heavily upon China.

Breathing in coal dust causes coalworker's pneumoconiosis which is known colloquially as "black lung", so-called because the coal dust literally turns the lungs black from their usual pink color. In the United States alone, it is estimated that 1500 former employees of the coal industry die every year from the effects of breathing in coal mine dust.

Around 10% of coal is ash, Coal ash is hazardous and toxic to human beings and other living things. Coal ash contains the radioactive elements uranium and thorium. Coal ash and other solid combustion byproducts are stored locally and escape in various ways that expose those living near coal plants to radiation and environmental toxics.

Huge amounts of coal ash and other waste is produced annually. In 2013, the US alone consumed on the order of 983 million short tonnes of coal per year. Use of coal on this scale generates hundreds of millions of tons of ash and other waste products every year. These include fly ash, bottom ash, and flue-gas desulfurization sludge, that contain mercury, uranium, thorium, arsenic, and other heavy metals, along with non-metals such as selenium.

The American Lung Association, the American Nurses' Association, and the Physicians for Social Responsibility released a report in 2009 which details in depth the detrimental impact of the coal industry on human health, including workers in the mines and individuals living in communities near plants burning coal as a power source. This report provides medical information regarding damage to the lungs, heart, and nervous system of Americans caused by the burning of coal as fuel. It details how the air pollution caused by the plume of coal smokestack emissions is a cause of asthma, strokes, reduced intelligence, artery blockages, heart attacks, congestive heart failure, cardiac arrhythmias, mercury poisoning, arterial occlusion, and lung cancer.

More recently, the Chicago School of Public Health released a largely similar report, echoing many of the same findings.

Though coal burning has increasingly been supplanted by less-toxic natural gas use in recent years, a 2010 study by the Clean Air Task Force still estimated that "air pollution from coal-fired power plants accounts for more than 13,000 premature deaths, 20,000 heart attacks, and 1.6 million lost workdays in the U.S. each year." The total monetary cost of these health impacts is over $100 billion annually.

The WHO classifies coal as a "dirty fuel" and encourages the movement away from such fuels towards cleaner alternatives.

Environmental Effects

Coal mining and coal fueling of power station and industrial processes can cause major environmental damage.

Water systems are affected by coal mining coal. For example, mining affects with groundwater and water table levels and acidity. Spills of fly ash, such as the Kingston Fossil Plant coal fly ash slurry spill, can also contaminate land and waterways, and destroy homes. Power stations that burn coal also consume large quantities of water. This can affect the flows of rivers, and has consequential impacts on other land uses.

One of the earliest known impacts of coal on the water cycle was acid rain. Approximately 75 Tg/S per year of sulfur dioxide (SO_2) is released from burning coal. After release, the sulfur dioxide is oxidized to gaseous H_2SO_2 which scatters solar radiation, hence its increase in the atmosphere exerts a cooling effect on climate. This beneficially masks some of the warming caused by increased greenhouse gases. However, the sulphur is precipitated out of the atmosphere as acid rain in a matter of weeks, whereas carbon dioxide remains in the atmosphere for hundreds of years. Release of SO_2 also contributes to the widespread acidification of ecosystems.

Disused coal mines can also cause issues. Subsidence can occur above tunnels, causing damage to infrastructure or cropland. Coal mining can also cause long lasting fires, and it has been estimated that around 1000 coal seam fires are burning at any given time. For example, there is a coal seam fire in Germany that has been burning since 1668, and is still burning in the 21st century.

Some environmental impacts are modest, such as dust nuisance. However, perhaps the largest and most long term effect of coal use is the release of carbon dioxide, a greenhouse gas that causes climate change and global warming, according to the IPCC and the EPA. Coal is the largest contributor to the human-made increase of CO_2 in the atmosphere.

The production of coke from coal produces ammonia, coal tar, and gaseous compounds as by-products which if discharged to land, air or waterways can act as environmental pollutants. The Whyalla steelworks is one example of a coke producing facility where liquid ammonia is discharged to the marine environment.

In 1999, world gross carbon dioxide emissions from coal usage were 8,666 million tonnes of carbon dioxide. In 2011, world gross emissions from coal usage were 14,416 million tonnes. For every megawatt-hour generated, coal-fired electric power generation emits around 2,000 pounds of carbon dioxide, which is almost double the approximately 1100 pounds of carbon dioxide released by a natural gas-fired electric plant. Because of this higher carbon efficiency of natural gas generation, as the market in the United States has changed to reduce coal and increase natural gas generation, carbon dioxide emissions may have fallen. Those measured in the first quarter of 2012 were the lowest of any recorded for the first quarter of any year since 1992. In 2013, the head of the UN climate agency advised that most of the world's coal reserves should be left in the ground to avoid catastrophic global warming.

Clean Coal Technology

"Clean" coal technology is a collection of technologies being developed to mitigate the environmental impact of coal energy generation. Those technologies are being developed to remove or reduce pollutant emissions to the atmosphere. Some of the techniques that would be used to accomplish this include chemically washing minerals and impurities from the coal, gasification, improved technology for treating flue gases to remove pollutants to increasingly stringent levels and at higher efficiency, carbon capture and storage technologies to capture the carbon dioxide from the flue gas and dewatering lower rank coals (brown coals) to improve the calorific value, and thus the efficiency of the conversion into electricity. Figures from the United States Environmental Protection Agency show that these technologies have made today's coal-based generating fleet 77 percent cleaner on the basis of regulated emissions per unit of energy produced.

Clean coal technology usually addresses atmospheric problems resulting from burning coal. Historically, the primary focus was on SO_2 and NO_x, the most important gases in causation of acid rain, and particulates which cause visible air pollution and deleterious effects on human health. More recent focus has been on carbon dioxide (due to its impact on global warming) and concern over toxic species such as mercury. Concerns exist regarding the economic viability of these technologies and the timeframe of delivery, potentially high hidden economic costs in terms of social and environmental damage, and the costs and viability of disposing of removed carbon and other toxic matter.

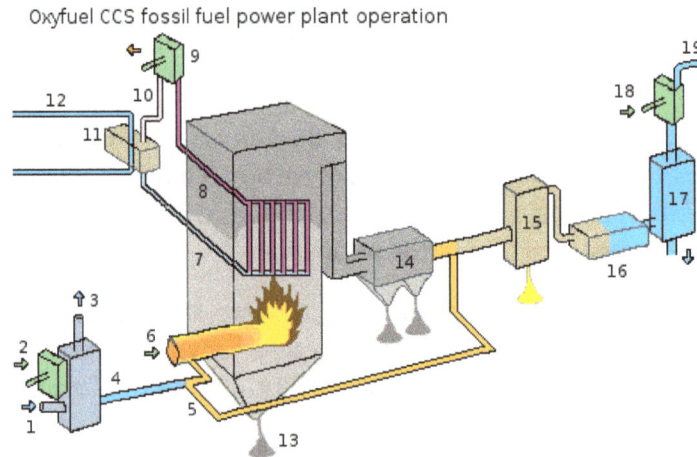

Oxyfuel CCS fossil fuel power plant operation

An oxyfuel CCS power plant operation processes the exhaust gases so as to separate the CO_2
so that it may be stored or sequestered

Several different technological methods are available for the purpose of carbon capture as demanded by the clean coal concept:

- Pre-combustion capture - This involves gasification of a feedstock (such as coal) to form synthesis gas, which may be shifted to produce a H_2 and CO_2-rich gas mixture, from which the CO_2 can be efficiently captured and separated, transported, and ultimately sequestered, This technology is usually associated with Integrated Gasification Combined Cycle process configurations.

- Post-combustion capture - This refers to capture of CO_2 from exhaust gases of combustion processes, typically using sorbents, solvents, or membrane separations to remove CO_2 from the bulk gases.

- Oxy-fuel combustion - Fossil fuels such as coal are burned in a mixture of recirculated flue gas and oxygen, rather than in air, which largely eliminates nitrogen from the flue gas enabling efficient, low-cost CO_2 capture.

The Kemper County IGCC Project, a 582 MW coal gasification-based power plant, will use pre-combustion capture of CO_2 to capture 65% of the CO_2 the plant produces, which will be utilized/geologically sequestered in enhanced oil recovery operations.

The Saskatchewan Government's Boundary Dam Integrated Carbon Capture and Sequestration Demonstration Project will use post-combustion, amine-based scrubber technology to capture 90% of the CO_2 emitted by Unit 3 of the power plant; this CO_2 will be pipelined to and utilized for enhanced oil recovery in the Weyburn oil fields.

An early example of a coal-based plant using (oxy-fuel) carbon-capture technology is Swedish company Vattenfall's Schwarze Pumpe power station located in Spremberg, Germany, built by German firm Siemens, which went on-line in September 2008. The facility captures CO_2 and acid rain producing pollutants, separates them, and compresses the CO_2 into a liquid. Plans are to inject the CO_2 into depleted natural gas fields or other geological formations. Vattenfall opines that this technology is considered not to be a final solution for CO_2 reduction in the atmosphere, but provides an achievable solution in the near term while more desirable alternative solutions to power generation can be made economically practical.

Bioremediation

The white rot fungus Trametes versicolor can grow on and metabolize naturally occurring coal. The bacteria Diplococcus has been found to degrade coal, raising its temperature.

Economic Aspects

Coal (by liquefaction technology) is one of the backstop resources that could limit escalation of oil prices and mitigate the effects of transportation energy shortage that will occur under peak oil. This is contingent on liquefaction production capacity becoming large enough to satiate the very large and growing demand for petroleum. Estimates of the cost of producing liquid fuels from coal suggest that domestic U.S. production of fuel from coal becomes cost-competitive with oil priced at around $35 per barrel, with the $35 being the break-even cost. With oil prices as low as around $40 per barrel in the U.S. as of December 2008, liquid coal lost some of its economic allure in the U.S., but will probably be re-vitalized, similar to oil sand projects, with an oil price around $70 per barrel.

In China, due to an increasing need for liquid energy in the transportation sector, coal liquefaction projects were given high priority even during periods of oil prices below $40 per barrel. This is probably because China prefers not to be dependent on foreign oil, instead utilizing its enormous domestic coal reserves. As oil prices were increasing during the first half of 2009, the coal liquefaction projects in China were again boosted, and these projects are profitable with an oil barrel price of $40.

China is the largest producer of coal in the world. It is the world's largest energy consumer, and relies on coal to supply 69% of its energy needs. An estimated 5 million people worked in China's coal-mining industry in 2007.

Coal pollution costs the EU €43 billion each year. Measures to cut air pollution may have beneficial long-term economic impacts for individuals.

Energy Density and Carbon Impact

The energy density of coal, i.e. its heating value, is roughly 24 megajoules per kilogram (approximately 6.7 kilowatt-hours per kg). For a coal power plant with a 40% efficiency, it takes an estimated 325 kg (717 lb) of coal to power a 100 W lightbulb for one year.

As of 2006, the average efficiency of electricity-generating power stations was 31%; in 2002, coal represented about 23% of total global energy supply, an equivalent of 3.4 billion tonnes of coal, of which 2.8 billion tonnes were used for electricity generation.

The US Energy Information Agency's 1999 report on CO_2 emissions for energy generation quotes an emission factor of 0.963 kg CO_2/kWh for coal power, compared to 0.881 kg CO_2/kWh (oil), or 0.569 kg CO_2/kWh (natural gas).

Underground Fires

Thousands of coal fires are burning around the world. Those burning underground can be difficult to locate and many cannot be extinguished. Fires can cause the ground above to subside, their combustion gases are dangerous to life, and breaking out to the surface can initiate surface wildfires. Coal seams can be set on fire by spontaneous combustion or contact with a mine fire or surface fire. Lightning strikes are an important source of ignition. The coal continues to burn slowly back into the seam until oxygen (air) can no longer reach the flame front. A grass fire in a coal area can set dozens of coal seams on fire. Coal fires in China burn an estimated 120 million tons of coal a year, emitting 360 million metric tons of CO_2, amounting to 2–3% of the annual worldwide production of CO_2 from fossil fuels. In Centralia, Pennsylvania (a borough located in the Coal Region of the United States), an exposed vein of anthracite ignited in 1962 due to a trash fire in the borough landfill, located in an abandoned anthracite strip mine pit. Attempts to extinguish the fire were unsuccessful, and it continues to burn underground to this day. The Australian Burning Mountain was originally believed to be a volcano, but the smoke and ash come from a coal fire that has been burning for some 6,000 years.

At Kuh i Malik in Yagnob Valley, Tajikistan, coal deposits have been burning for thousands of years, creating vast underground labyrinths full of unique minerals, some of them very beautiful. Local people once used this method to mine ammoniac. This place has been well-known since the time of Herodotus, but European geographers misinterpreted the Ancient Greek descriptions as the evidence of active volcanism in Turkestan (up to the 19th century, when the Russian army invaded the area).

The reddish siltstone rock that caps many ridges and buttes in the Powder River Basin in Wyoming and in western North Dakota is called *porcelanite*, which resembles the coal burning waste "clinker" or volcanic "scoria". Clinker is rock that has been fused by the natural burning of coal. In the Powder River Basin approximately 27 to 54 billion tons of coal burned within the past three million years. Wild coal fires in the area were reported by the Lewis and Clark Expedition as well as explorers and settlers in the area.

Production Trends

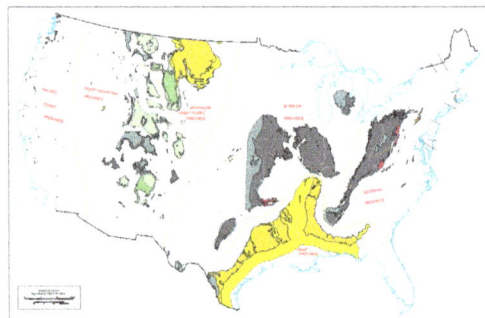

Continental United States coal regions

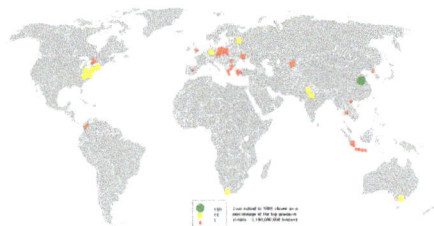

Coal output in 2005

A coal mine in Wyoming, United States. The United States has the world's largest coal reserves.

In 2006, China was the top producer of coal with 38% share followed by the United States and India, according to the British Geological Survey. As of 2012 coal production in the United States was falling at the rate of 7% annually with many power plants using coal shut down or converted to natural gas; however, some of the reduced domestic demand was taken up by increased exports with five coal export terminals being proposed in the Pacific Northwest to export coal from the Powder River Basin to China and other Asian markets; however, as of 2013, environmental opposition was increasing. High-sulfur coal mined in Illinois which was unsaleable in the United States found a ready market in Asia as exports reached 13 million tons in 2012.

World Coal Reserves

The 948 billion short tons of recoverable coal reserves estimated by the Energy Information Administration are equal to about 4,196 BBOE (billion barrels of oil equivalent). The amount of coal burned during 2007 was estimated at 7.075 billion short tons, or 133.179 quadrillion BTU's. This is an average of 18.8 million BTU per short ton. In terms of heat content, this is about 57,000,000 barrels (9,100,000 m³) of oil equivalent per day. By comparison in 2007, natural gas provided 51,000,000 barrels (8,100,000 m³) of oil equivalent per day, while oil provided 85,800,000 barrels (13,640,000 m³) per day.

British Petroleum, in its 2007 report, estimated at 2006 end that there were 147 years reserves-to-production ratio based on *proven* coal reserves worldwide. This figure only includes reserves classified as "proven"; exploration drilling programs by mining companies, particularly in under-explored areas, are continually providing new reserves. In many cases, companies are aware of coal deposits that have not been sufficiently drilled to qualify as "proven". However, some nations haven't updated their information and assume reserves remain at the same levels even with withdrawals.

Of the three fossil fuels, coal has the most widely distributed reserves; coal is mined in over 100 countries, and on all continents except Antarctica. The largest reserves are found in the United States, Russia, China, Australia and India. Note the table below.

Proved recoverable coal reserves at end-2008 or 2011 (million tons (teragrams))						
Country	Anthracite & Bituminous	SubBituminous	Lignite	Total	Percentage of World Total	Year
United States	108,501	98,618	30,176	237,295	22.6	2011
Russia	49,088	97,472	10,450	157,010	14.4	2011
China	62,200	33,700	18,600	114,500	12.6	2011
Australia	37,100	2,100	37,200	76,400	8.9	2011
India	56,100	0	4,500	60,600	7.0	2011
Germany	99	0	40,600	40,699	4.7	
Ukraine	15,351	16,577	1,945	33,873	3.9	
Kazakhstan	21,500	0	12,100	33,600	3.9	
South Africa	30,156	0	0	30,156	3.5	
Serbia	9	361	13,400	13,770	1.6	
Colombia	6,366	380	0	6,746	0.8	
Canada	3,474	872	2,236	6,528	0.8	
Poland	4,338	0	1,371	5,709	0.7	
Indonesia	1,520	2,904	1,105	5,529	0.6	
Brazil	0	4,559	0	4,559	0.5	
Greece	0	0	3,020	3,020	0.4	
Bosnia and Herzegovina	484	0	2,369	2,853	0.3	
Mongolia	1,170	0	1,350	2,520	0.3	
Bulgaria	2	190	2,174	2,366	0.3	
Pakistan	0	166	1,904	2,070	0.3	
Turkey	529	0	1,814	2,343	0.3	
Uzbekistan	47	0	1,853	1,900	0.2	
Hungary	13	439	1,208	1,660	0.2	
Thailand	0	0	1,239	1,239	0.1	
Mexico	860	300	51	1,211	0.1	
Iran	1,203	0	0	1,203	0.1	
Czech Republic	192	0	908	1,100	0.1	
Kyrgyzstan	0	0	812	812	0.1	
Albania	0	0	794	794	0.1	
North Korea	300	300	0	600	0.1	
New Zealand	33	205	333-7,000	571–15,000	0.1	

Proved recoverable coal reserves at end-2008 or 2011 (million tons (teragrams))						
Country	Anthracite & Bituminous	SubBituminous	Lignite	Total	Percentage of World Total	Year
Spain	200	300	30	530	0.1	
Laos	4	0	499	503	0.1	
Zimbabwe	502	0	0	502	0.1	
Argentina	0	0	550	550	0.1	2011
All others	3,421	1,346	846	5,613	0.7	
World Total	403,197	287,333	201,000	891,530	100	2011

Major Coal Producers

The reserve life is an estimate based only on current production levels and proved reserves level for the countries shown, and makes no assumptions of future production or even current production trends. Countries with annual production higher than 100 million tonnes are shown. For comparison, data for the European Union is also shown. Shares are based on data expressed in tonnes oil equivalent.

Production of Coal by Country and year (million tonnes)											
Country	2003	2004	2005	2006	2007	2008	2009	2010	2011	Share	Reserve Life (years)
China	1834.9	2122.6	2349.5	2528.6	2691.6	2802.0	2973.0	3235.0	3520.0	49.5%	35
United States	972.3	1008.9	1026.5	1054.8	1040.2	1063.0	975.2	983.7	992.8	14.1%	239
India	375.4	407.7	428.4	449.2	478.4	515.9	556.0	573.8	588.5	5.6%	103
European Union	637.2	627.6	607.4	595.1	592.3	563.6	538.4	535.7	576.1	4.2%	97
Australia	350.4	364.3	375.4	382.2	392.7	399.2	413.2	424.0	415.5	5.8%	184
Russia	276.7	281.7	298.3	309.9	313.5	328.6	301.3	321.6	333.5	4.0%	471
Indonesia	114.3	132.4	152.7	193.8	216.9	240.2	256.2	275.2	324.9	5.1%	17
South Africa	237.9	243.4	244.4	244.8	247.7	252.6	250.6	254.3	255.1	3.6%	118
Germany	204.9	207.8	202.8	197.1	201.9	192.4	183.7	182.3	188.6	1.1%	216
Poland	163.8	162.4	159.5	156.1	145.9	144.0	135.2	133.2	139.2	1.4%	41
Kazakhstan	84.9	86.9	86.6	96.2	97.8	111.1	100.9	110.9	115.9	1.5%	290
World Total	5,301.3	5,716.0	6,035.3	6,342.0	6,573.3	6,795.0	6,880.8	7,254.6	7,695.4	100%	112

Major Coal Consumers

Countries with annual consumption higher than 20 million tonnes are shown.

Consumption of Coal by Country and year (million short tons)					
Country	2008	2009	2010	2011	Share
China	2,966	3,188	3,695	4,053	50.7%
United States	1,121	997	1,048	1,003	12.5%
India	641	705	722	788	9.9%
Russia	250	204	256	262	3.3%
Germany	268	248	256	256	3.3%
South Africa	215	204	206	210	2.6%
Japan	204	181	206	202	2.5%
Poland	149	151	149	162	2.0%
World Total	7,327	7,318	7,994	N/A	100%

Major Coal Exporters

Countries with annual gross export higher than 10 million tonnes are shown. In terms of net export the largest exporters are still Australia (328.1 millions tonnes), Indonesia (316.2) and Russia (100.2).

Exports of Coal by Country and year (million short tons)											
Country	2003	2004	2005	2006	2007	2008	2009	2010	2011	2012	Share
Indonesia	107.8	131.4	142.0	192.2	221.9	228.2	261.4	316.2	331.4	421.8	29.8%
Australia	238.1	247.6	255.0	255.0	268.5	278.0	288.5	328.1	313.6	332.4	23.5%
Russia	41.0	55.7	98.6	103.4	112.2	115.4	130.9	122.1	140.1	150.7	10.7%
United States	43.0	48.0	51.7	51.2	60.6	83.5	60.4	83.2	108.2	126.7	8.7%
Colombia	50.4	56.4	59.2	68.3	74.5	74.7	75.7	76.4	89.0	92.2	6.5%
South Africa	78.7	74.9	78.8	75.8	72.6	68.2	73.8	76.7	75.8	82.0	5.8%
Canada	27.7	28.8	31.2	31.2	33.4	36.5	31.9	36.9	37.6	38.8	2.7%
Kazakhstan	30.3	27.4	28.3	30.5	32.8	47.6	33.0	36.3	33.5	35.2	2.5%
Mongolia	0.5	1.7	2.3	2.5	3.4	4.4	7.7	18.3	26.1	24.3	1.7%
Vietnam	6.9	11.7	19.8	23.5	35.1	21.3	28.2	24.7	19.7	21.1	1.5%
China	103.4	95.5	93.1	85.6	75.4	68.8	25.2	22.7	27.5	15.2	1.1%
Poland	28.0	27.5	26.5	25.4	20.1	16.1	14.6	18.1	15.0	14.9	1.0%
Total World	713.9	764.0	936.0	1,000.6	1,073.4	1,087.3	1,090.8	1,212.8	1,286.7	1,413.9	100%

Major Coal Importers

Countries with annual gross import higher than 20 million tonnes are shown. In terms of net import the largest importers are still Japan (206.0 millions tonnes), China (172.4) and South Korea (125.8).

Imports of Coal by Country and year (million short tons)						
Country	**2006**	**2007**	**2008**	**2009**	**2010**	**Share**
Japan	199.7	209.0	206.0	182.1	206.7	17.5%
China	42.0	56.2	44.5	151.9	195.1	16.6%
South Korea	84.1	94.1	107.1	109.9	125.8	10.7%
India	52.7	29.6	70.9	76.7	101.6	8.6%
Taiwan	69.1	72.5	70.9	64.6	71.1	6.0%
Germany	50.6	56.2	55.7	45.9	55.1	4.7%
Turkey	22.9	25.8	21.7	22.7	30.0	2.5%
United Kingdom	56.8	48.9	49.2	42.2	29.3	2.5%
Italy	27.9	28.0	27.9	20.9	23.7	1.9%
Netherlands	25.7	29.3	23.5	22.1	22.8	1.9%
Russia	28.8	26.3	34.6	26.8	21.8	1.9%
France	24.1	22.1	24.9	18.3	20.8	1.8%
United States	40.3	38.8	37.8	23.1	20.6	1.8%
Total	**991.8**	**1,056.5**	**1,063.2**	**1,039.8**	**1,178.1**	**100%**

Cultural Usage

Coal is the official state mineral of Kentucky. and the official state rock of Utah; both U.S. states have a historic link to coal mining.

Some cultures hold that children who misbehave will receive only a lump of coal from Santa Claus for Christmas in their christmas stockings instead of presents.

It is also customary and considered lucky in Scotland and the North of England to give coal as a gift on New Year's Day. This occurs as part of First-Footing and represents warmth for the year to come.

Oil Shale

Oil shale, also known as kerogen shale, is an organic-rich fine-grained sedimentary rock containing kerogen (a solid mixture of organic chemical compounds) from which liquid hydrocarbons called shale oil (not to be confused with tight oil—crude oil occurring naturally in shales) can be produced. Shale oil is a substitute for conventional crude oil; however, extracting shale oil from oil shale is more costly than the production of conventional crude oil both financially and in terms of its environmental impact. Deposits of oil shale occur around the world, including major deposits in the United States. Estimates of global deposits range from 4.8 to 5 trillion barrels (760×10^9 to 790×10^9 m^3) of oil in place.

Heating oil shale to a sufficiently high temperature causes the chemical process of pyrolysis to yield a vapor. Upon cooling the vapor, the liquid shale oil—an unconventional oil—is separated from combustible oil-shale gas (the term *shale gas* can also refer to gas occurring naturally in shales). Oil shale can also be burned directly in furnaces as a low-grade fuel for power generation and district heating or used as a raw material in chemical and construction-materials processing.

Oil shale gains attention as a potential abundant source of oil whenever the price of crude oil rises. At the same time, oil-shale mining and processing raise a number of environmental concerns, such as land use, waste disposal, water use, waste-water management, greenhouse-gas emissions and air pollution. Estonia and China have well-established oil shale industries, and Brazil, Germany, and Russia also utilize oil shale.

General composition of oil shales constitutes inorganic matrix, bitumens, and kerogen. Oil shales differ from oil-*bearing* shales, shale deposits that contain petroleum (tight oil) that is sometimes produced from drilled wells. Examples of oil-*bearing* shales are the Bakken Formation, Pierre Shale, Niobrara Formation, and Eagle Ford Formation.

Geology

Outcrop of Ordovician oil shale (kukersite), northern Estonia

Oil shale, an organic-rich sedimentary rock, belongs to the group of sapropel fuels. It does not have a definite geological definition nor a specific chemical formula, and its seams do not always have discrete boundaries. Oil shales vary considerably in their mineral content, chemical composition, age, type of kerogen, and depositional history and not all oil shales would necessarily be classified as shales in the strict sense. According to the petrologist Adrian C. Hutton of the University of Wollongong, oil shales are not "geological nor geochemically distinctive rock but rather 'economic' term." Their common feature is low solubility in low-boiling organic solvents and generation of liquid organic products on thermal decomposition.

Oil shale differs from bitumen-impregnated rocks (oil sands and petroleum reservoir rocks), humic coals and carbonaceous shale. While oil sands do originate from the biodegradation of oil, heat and pressure have not (yet) transformed the kerogen in oil shale into petroleum, that means that its maturation does not exceed early mesocatagenetic.

General composition of oil shales constitutes inorganic matrix, bitumens, and kerogen. While the bitumen portion of oil shales is soluble in carbon disulfide, kerogen portion is insoluble in carbon disulfide and can contain iron, vanadium, nickel, molybdenum, and uranium. Oil shale contains a lower percentage of organic matter than coal. In commercial grades of oil shale the ratio of organic matter to mineral matter lies approximately between 0.75:5 and 1.5:5. At the same time, the organic matter in oil shale has an atomic ratio of hydrogen to carbon (H/C) approximately 1.2 to 1.8 times lower than for crude oil and about 1.5 to 3 times higher than for coals. The organic components of oil shale derive from a variety of organisms, such as the remains of algae, spores, pollen, plant cuticles and corky fragments of herbaceous and woody plants, and cellular debris from other aquatic and land plants. Some deposits contain significant fossils; Germany's Messel Pit has the status of a Unesco World Heritage Site. The mineral matter in oil shale includes various fine-grained silicates and carbonates. Inorganic matrix can contain quartz, feldspars, clays (mainly illite and chlorite), carbonates (calcite and dolomites), pyrite and some other minerals.

Geologists can classify oil shales on the basis of their composition as carbonate-rich shales, siliceous shales, or cannel shales.

Another classification, known as the van Krevelen diagram, assigns kerogen types, depending on the hydrogen, carbon, and oxygen content of oil shales' original organic matter. The most commonly used classification of oil shales, developed between 1987 and 1991 by Adrian C. Hutton, adapts petrographic terms from coal terminology. This classification designates oil shales as terrestrial, lacustrine (lake-bottom-deposited), or marine (ocean bottom-deposited), based on the environment of the initial biomass deposit. Known oil shales are predominantly aquatic (marine, lacustrine) origin. Hutton's classification scheme has proven useful in estimating the yield and composition of the extracted oil.

Resource

Fossils in Ordovician oil shale (kukersite), northern Estonia

As source rocks for most conventional oil reservoirs, oil shale deposits are found in all world oil provinces, although most of them are too deep to be exploited economically. As with all oil and gas resources, analysts distinguish between oil shale resources and oil shale reserves. "Resources" refers to all oil shale deposits, while "reserves", represents those deposits from which producers can extract oil shale economically using existing technology. Since extraction technologies develop continuously, planners can only estimate the amount of recoverable kerogen. Although resources of oil shale occur in many countries, only 33 countries possess known deposits of possible economic value . Well-explored deposits, potentially classifiable as reserves, include the Green River deposits in the western United States, the Tertiary deposits in Queensland, Australia, deposits in Sweden and Estonia, the El-Lajjun deposit in Jordan, and deposits in France, Germany, Brazil, China, southern Mongolia and Russia. These deposits have given rise to expectations of yielding at least 40 liters of shale oil per tonne of oil shale, using the Fischer Assay.

A 2008 estimate set the total world resources of oil shale at 689 gigatons — equivalent to yield of 4.8 trillion barrels (760 billion cubic metres) of shale oil, with the largest reserves in the United States, which is thought to have 3.7 trillion barrels (590 billion cubic metres), though only a part of it is recoverable. According to the 2010 World Energy Outlook by the International Energy Agency, the world oil shale resources may be equivalent of more than 5 trillion barrels (790 billion cubic metres) of oil in place of which more than 1 trillion barrels (160 billion cubic metres) may be technically recoverable. For comparison, the world's proven conventional oil reserves were estimated at 1.317 trillion barrels (209.4×10^9 m³), as of 1 January 2007. The largest deposits in the world occur in the United States in the Green River Formation, which covers portions of Colorado, Utah, and Wyoming; about 70% of this resource lies on land owned or managed by the United States federal government. Deposits in the United States constitute 62% of world resources; together, the United States, Russia and Brazil account for 86% of the world's resources in terms of shale-oil content. These figures remain tentative, with exploration or analysis of several deposits still outstanding. Professor Alan R. Carroll of University of Wisconsin–Madison regards the Upper Permian lacustrine oil-shale deposits of northwest China, absent from previous global oil shale assessments, as comparable in size to the Green River Formation.

History

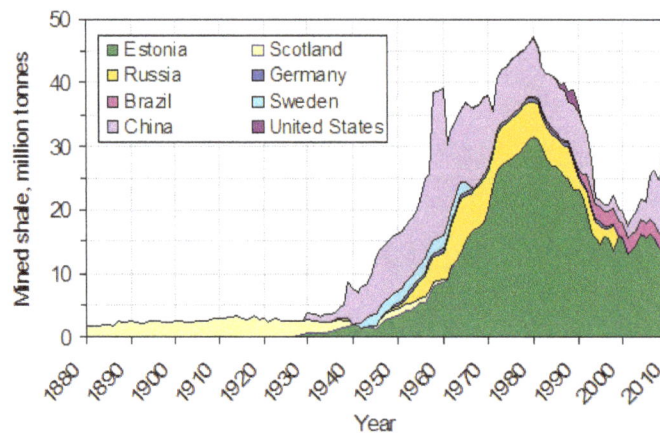

Production of oil shale in millions of metric tons, from 1880 to 2010. Source: Pierre Allix, Alan K. Burnham.

Humans have used oil shale as a fuel since prehistoric times, since it generally burns without any processing. Britons of the Iron Age also used to polish it and form it into ornaments. The first patent for extracting oil from oil shale was British Crown Patent 330 granted in 1694 to three persons named Martin Eele, Thomas Hancock and William Portlock who had "found a way to extract and make great quantities of pitch, tarr, and oyle out of a sort of stone."

Autun oil shale mines

Modern industrial mining of oil shale began in 1837 in Autun, France, followed by exploitation in Scotland, Germany, and several other countries.

Operations during the 19th century focused on the production of kerosene, lamp oil, and paraffin; these products helped supply the growing demand for lighting that arose during the Industrial Revolution. Fuel oil, lubricating oil and grease, and ammonium sulfate were also produced. The European oil-shale industry expanded immediately before World War I due to limited access to conventional petroleum resources and to the mass production of automobiles and trucks, which accompanied an increase in gasoline consumption.

Although the Estonian and Chinese oil-shale industries continued to grow after World War II, most other countries abandoned their projects due to high processing costs and the availability of cheaper petroleum. Following the 1973 oil crisis, world production of oil shale reached a peak of 46 million tonnes in 1980 before falling to about 16 million tonnes in 2000, due to competition from cheap conventional petroleum in the 1980s.

On 2 May 1982, known in some circles as "Black Sunday", Exxon canceled its US$5 billion Colony Shale Oil Project near Parachute, Colorado because of low oil-prices and increased expenses, laying off more than 2,000 workers and leaving a trail of home-foreclosures and small-business bankruptcies. In 1986, President Ronald Reagan signed into law the Consolidated Omnibus Budget Reconciliation Act of 1985 which among other things abolished the United States' Synthetic Liquid Fuels Program.

The global oil-shale industry began to revive at the beginning of the 21st century. In 2003, an oil-shale development program restarted in the United States. Authorities introduced a commercial leasing program permitting the extraction of oil shale and oil sands on federal lands in 2005, in accordance with the Energy Policy Act of 2005.

Industry

As of 2008, industry uses oil shale in Brazil, China, Estonia and to some extent in Germany, and Russia. Several additional countries started assessing their reserves or had built experimental production plants, while others had phased out their oil shale industry. Oil shale serves for oil production in Estonia, Brazil, and China; for power generation in Estonia, China, and Germany; for cement production in Estonia, Germany, and China; and for use in chemical industries in China, Estonia, and Russia.

Shell's experimental *in-situ* oil-shale facility, Piceance Basin, Colorado, USA

As of 2009, 80% of oil shale used globally is extracted in Estonia, mainly due to the Oil-shale-fired power plants. Oil-shale-fired power plants occur in Estonia, which has an installed capacity of 2,967 megawatts (MW), China (12 MW), and Germany (9.9 MW). Israel, Romania and Russia have in the past run power plants fired by oil shale, but have shut them down or switched to other fuel sources such as natural gas. Jordan and Egypt plan to construct power plants fired by oil shale, while Canada and Turkey plan to burn oil shale along with coal for power generation. Oil shale serves as the main fuel for power generation only in Estonia, where the oil-shale-fired Narva Power Plants accounted for 95% of country's electrical generation in 2005.

According to the World Energy Council, in 2008 the total production of shale oil from oil shale was 930,000 tonnes, equal to 17,700 barrels per day (2,810 m³/d), of which China produced 375,000 tonnes, Estonia 355,000 tonnes, and Brazil 200,000 tonnes. In comparison, production of the conventional oil and natural gas liquids in 2008 amounted 3.95 billion tonnes or 82.1 million barrels per day (13.1×10^6 m³/d).

Extraction and Processing

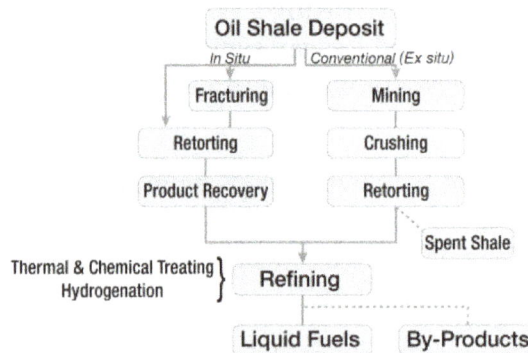

Overview of shale oil extraction.

Most exploitation of oil shale involves mining followed by shipping elsewhere, after which one can burn the shale directly to generate electricity, or undertake further processing. The most common methods of surface mining involve open pit mining and strip mining. These procedures remove most of the overlying material to expose the deposits of oil shale, and become practical when the deposits occur near the surface. Underground mining of oil shale, which removes less of the overlying material, employs the room-and-pillar method.

The extraction of the useful components of oil shale usually takes place above ground (*ex-situ* processing), although several newer technologies perform this underground (on-site or *in-situ* processing). In either case, the chemical process of pyrolysis converts the kerogen in the oil shale to shale oil (synthetic crude oil) and oil shale gas. Most conversion technologies involve heating shale in the absence of oxygen to a temperature at which kerogen decomposes (pyrolyses) into gas, condensable oil, and a solid residue. This usually takes place between 450 °C (842 °F) and 500 °C (932 °F). The process of decomposition begins at relatively low temperatures (300 °C or 572 °F), but proceeds more rapidly and more completely at higher temperatures.

In-situ processing involves heating the oil shale underground. Such technologies can potentially extract more oil from a given area of land than *ex-situ* processes, since they can access the material at greater depths than surface mines can. Several companies have patented methods for *in-situ* retorting. However, most of these methods remain in the experimental phase. One can distinguish *true in-situ* processes (TIS) and *modified in-situ* processes (MIS). *True in-situ* processes do not involve mining the oil shale. *Modified in-situ* processes involve removing part of the oil shale and bringing it to the surface for modified *in-situ* retorting in order to create permeability for gas flow in a rubble chimney. Explosives rubblize the oil-shale deposit.

Hundreds of patents for oil shale retorting technologies exist; however, only a few dozen have undergone testing. By 2006, only four technologies remained in commercial use: Kiviter, Galoter, Fushun, and Petrosix.

Applications and Products

Industry can use oil shale as a fuel for thermal power-plants, burning it (like coal) to drive steam turbines; some of these plants employ the resulting heat for district heating of homes and businesses. In addition to its use as a fuel, oil shale may also serve in the production of specialty carbon fibers, adsorbent carbons, carbon black, phenols, resins, glues, tanning agents, mastic, road bitumen, cement, bricks, construction and decorative blocks, soil-additives, fertilizers, rock-wool insulation, glass, and pharmaceutical products. However, oil shale use for production of these items remains small or only in its experimental stages. Some oil shales yield sulfur, ammonia, alumina, soda ash, uranium, and nahcolite as shale-oil extraction byproducts. Between 1946 and 1952, a marine type of *Dictyonema* shale served for uranium production in Sillamäe, Estonia, and between 1950 and 1989 Sweden used alum shale for the same purposes. Oil shale gas has served as a substitute for natural gas, but as of 2009, producing oil shale gas as a natural-gas substitute remained economically infeasible.

The shale oil derived from oil shale does not directly substitute for crude oil in all applications. It may contain higher concentrations of olefins, oxygen, and nitrogen than conventional crude oil. Some shale oils may have higher sulfur or arsenic content. By comparison with West Texas Intermediate, the benchmark standard for crude oil in the futures-contract market, the Green River shale oil sulfur content ranges from near 0% to 4.9% (in average 0.76%), where West Texas Intermediate's sulfur content has a maximum of 0.42%. The sulfur content in shale oil from Jordan's oil shales may rise even up to 9.5%. The arsenic content, for example, becomes an issue for Green River formation oil shale. The higher concentrations of these materials means that the oil must undergo considerable upgrading (hydrotreating) before serving as oil-refinery feedstock. Above-ground retorting processes tended to yield a lower API gravity shale oil than the *in situ* processes. Shale oil serves best for producing middle-distillates such as kerosene, jet fuel, and diesel fuel. Worldwide demand for these middle distillates, particularly for diesel fuels, increased rapidly in the 1990s and 2000s. However, appropriate refining processes equivalent to hydrocracking can transform shale oil into a lighter-range hydrocarbon (gasoline).

Economics

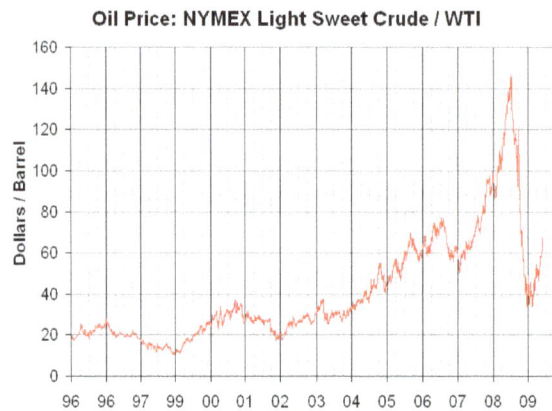

Oil Price: NYMEX Light Sweet Crude / WTI

NYMEX light-sweet crude oil prices 1996–2009 (not adjusted for inflation)

The amount of economically recoverable oil shale is unknown. The various attempts to develop oil shale deposits have succeeded only when the cost of shale-oil production in a given region comes in below the price of crude oil or its other substitutes. According to a survey conducted by the RAND Corporation, the cost of producing a barrel of oil at a surface retorting complex in the United States (comprising a mine, retorting plant, upgrading plant, supporting utilities, and spent shale reclamation), would range between US\$70–95 (\$440–600/m^3, adjusted to 2005 values). This estimate considers varying levels of kerogen quality and extraction efficiency. In order to run a profitable operation, the price of crude oil would need to remain above these levels. The analysis also discusses the expectation that processing costs would drop after the establishment of the complex. The hypothetical unit would see a cost reduction of 35–70% after producing its first 500 million barrels (79×10^6 m^3). Assuming an increase in output of 25 thousand barrels per day (4.0×10^3 m^3/d) during each year after the start of commercial production, RAND predicts the costs would decline to \$35–48 per barrel (\$220–300/m^3) within 12 years. After achieving the milestone of 1 billion barrels (160×10^6 m^3), its costs would decline further to \$30–40 per barrel (\$190–250/m^3). Some commentators compare the proposed American oil-shale industry to the Athabasca oil-sands industry (the latter enterprise generated over 1 million barrels (160,000 m^3) of oil per day in late 2007), stating that "the first-generation facility is the hardest, both technically and economically".

In 2005, Royal Dutch Shell announced that its *in-situ* process could become competitive for oil prices over \$30 per barrel (\$190/m^3). A 2004 report by the United States Department of Energy stated that both the Shell technology and technology used in the Stuart Oil Shale Project could be competitive at prices above $25 per barrel, and that the Viru Keemia Grupp expected full-scale production to be economical at prices above $18 per barrel ($130/m³).

To increase efficiency when retorting oil shale, researchers have proposed and tested several co-pyrolysis processes.

A 1972 publication in the journal *Pétrole Informations* (ISSN 0755-561X) compared shale-based oil production unfavorably with coal liquefaction. The article portrayed coal liquefaction as less expensive, generating more oil, and creating fewer environmental impacts than extraction from oil shale. It cited a conversion ratio of 650 liters (170 U.S. gal; 140 imp gal) of oil per one ton of coal, as against 150 liters (40 U.S. gal; 33 imp gal) of shale oil per one ton of oil shale.

A critical measure of the viability of oil shale as an energy source lies in the ratio of the energy produced by the shale to the energy used in its mining and processing, a ratio known as "Energy Returned on Energy Invested" (EROEI). A 1984 study estimated the EROEI of the various known oil-shale deposits as varying between 0.7–13.3 although known oil-shale extraction development projects assert an EROEI between 3 and 10. According to the World Energy Outlook 2010, the EROEI of *ex-situ* processing is typically 4 to 5 while of *in-situ* processing it may be even as low as 2. However, according to the IEA most of used energy can be provided by burning the spent shale or oil-shale gas.

The water needed in the oil shale retorting process offers an additional economic consideration: this may pose a problem in areas with water scarcity.

Environmental Considerations

Mining oil shale involves a number of environmental impacts, more pronounced in surface mining than in underground mining. These include acid drainage induced by the sudden rapid exposure and subsequent oxidation of formerly buried materials, the introduction of metals including mercury into surface-water and groundwater, increased erosion, sulfur-gas emissions, and air pollution caused by the production of particulates during processing, transport, and support activities. In 2002, about 97% of air pollution, 86% of total waste and 23% of water pollution in Estonia came from the power industry, which uses oil shale as the main resource for its power production.

Oil-shale extraction can damage the biological and recreational value of land and the ecosystem in the mining area. Combustion and thermal processing generate waste material. In addition, the atmospheric emissions from oil shale processing and combustion include carbon dioxide, a greenhouse gas. Environmentalists oppose production and usage of oil shale, as it creates even more greenhouse gases than conventional fossil fuels. Section 526 of the *Energy Independence And Security Act* prohibits United States government agencies from buying oil produced by processes that produce more greenhouse gas emissions than would traditional petroleum. Experimental *in situ* conversion processes and carbon capture and storage technologies may reduce some of these concerns in the future, but at the same time they may cause other problems, including groundwater pollution. Among the water contaminants commonly associated with oil shale processing are oxygen and nitrogen heterocyclic hydrocarbons. Commonly detected examples include quinoline derivatives, pyridine, and various alkyl homologues of pyridine (picoline, lutidine).

Some have expressed concerns over the oil shale industry's use of water. In 2002, the oil shale-fired power industry used 91% of the water consumed in Estonia. Depending on technology, above-ground retorting uses between one and five barrels of water per barrel of produced shale-oil. A 2008 programmatic environmental impact statement issued by the US Bureau of Land Management stated that surface mining and retort operations produce 2 to 10 U.S. gallons (7.6 to 37.9 l; 1.7 to 8.3 imp gal) of waste water per 1 short ton (0.91 t) of processed oil shale. *In situ* processing, according to one estimate, uses about one-tenth as much water.

Water concerns become particularly sensitive issues in arid regions, such as the western US and Israel's Negev Desert, where plans exist to expand oil-shale extraction despite a water shortage.

Environmental activists, including members of Greenpeace, have organized strong protests against the oil shale industry. In one result, Queensland Energy Resources put the proposed Stuart Oil Shale Project in Australia on hold in 2004.

Extraterrestrial Oil Shale

Some comets contain "massive amounts of an organic material almost identical to high grade oil shale," the equivalent of cubic kilometers of such mixed with other material; for instance, corresponding hydrocarbons were detected in a probe fly-by through the tail of Comet Halley during 1986.

Clay

Gay Head cliffs in Martha's Vineyard consist almost entirely of clay.

Clay is a fine-grained natural rock or soil material that combines one or more clay minerals with traces of metal oxides and organic matter. Clays are plastic due to their water content and become hard, brittle and non–plastic upon drying or firing. Geologic clay deposits are mostly composed of phyllosilicate minerals containing variable amounts of water trapped in the mineral structure. Depending on the soil's content in which it is found, clay can appear in various colours from white to dull gray or brown to deep orange-red.

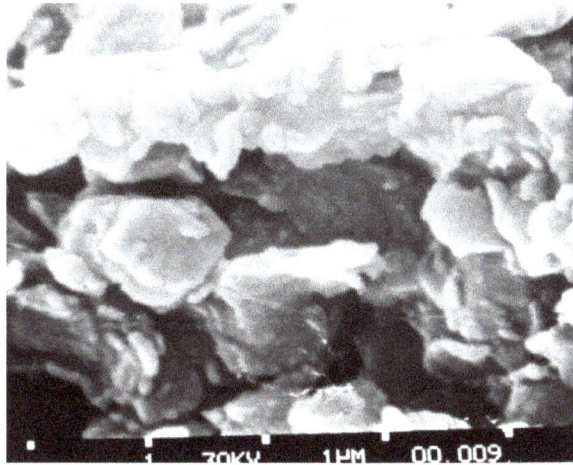

Electron microscope photograph of smectite clay – magnification 23,500

Clays are distinguished from other fine-grained soils by differences in size and mineralogy. Silts, which are fine-grained soils that do not include clay minerals, tend to have larger particle sizes than clays. There is, however, some overlap in particle size and other physical properties, and many naturally occurring deposits include both silts and clay. The distinction between silt and clay varies by discipline. Geologists and soil scientists usually consider the separation to occur at a particle size of 2 μm (clays being finer than silts), sedimentologists often use 4–5 μm, and colloid chemists use 1 μm. Geotechnical engineers distinguish between silts and clays based on the plasticity properties of the soil, as measured by the soils' Atterberg limits. ISO 14688 grades clay particles as being smaller than 2 μm and silt particles as being larger.

Quaternary clay in Estonia

Formation

Clay minerals typically form over long periods of time as a result of the gradual chemical weathering of rocks, usually silicate-bearing, by low concentrations of carbonic acid and other diluted solvents. These solvents, usually acidic, migrate through the weathering rock after leaching through upper weathered layers. In addition to the weathering process, some clay minerals are formed through hydrothermal activity. There are two types of clay deposits: primary and secondary. Primary clays form as residual deposits in soil and remain at the site of formation. Secondary clays are clays that have been transported from their original location by water erosion and deposited in a new sedimentary deposit. Clay deposits are typically associated with very low energy depositional environments such as large lakes and marine basins.

Deforestation for clay extraction in Rio de Janeiro, Brazil. The picture is of Morro da Covanca, Jacarepaguá.

Grouping

Depending on the academic source, there are three or four main groups of clays: kaolinite, montmorillonite-smectite, illite, and chlorite. Chlorites are not always considered a clay, sometimes being classified as a separate group within the phyllosilicates. There are approximately 30 different types of "pure" clays in these categories, but most "natural" clay deposits are mixtures of these different types, along with other weathered minerals.

Varve (or *varved clay*) is clay with visible annual layers, which are formed by seasonal deposition of those layers and are marked by differences in erosion and organic content. This type of deposit is common in former glacial lakes. When fine sediments are delivered into the calm waters of these glacial lake basins away from the shoreline, they settle to the lake bed. The resulting seasonal layering is preserved in an even distribution of clay sediment banding.

Quick clay is a unique type of marine clay indigenous to the glaciated terrains of Norway, Canada, Northern Ireland, and Sweden. It is a highly sensitive clay, prone to liquefaction, which has been involved in several deadly landslides.

Historical and Modern Uses

Clay layers in a construction site. Dry clay is normally much more stable than sand with regard to excavations.

Clays exhibit plasticity when mixed with water in certain proportions. When dry, clay becomes firm and when fired in a kiln, permanent physical and chemical changes occur. These changes convert the clay into a ceramic material. Because of these properties, clay is used for making pottery, both utilitarian and decorative, and construction products, such as bricks, wall and floor tiles. Different types of clay, when used with different minerals and firing conditions, are used to produce earthenware, stoneware, and porcelain. Prehistoric humans discovered the useful properties of clay. Some of the earliest pottery shards recovered are from central Honshu, Japan. They are associated with the Jomon culture and deposits they were recovered from have been dated to around 14,000 BC.

Bottle stopper made of clay, 14th century

Clay tablets were the first known writing medium. Scribes wrote by inscribing them with cuneiform script using a blunt reed called a stylus. Purpose-made clay balls were also used as sling ammunition.

Clays sintered in fire were the first form of ceramic. Bricks, cooking pots, art objects, dishware, and even musical instruments such as the ocarina can all be shaped from clay before being fired. Clay is also used in many industrial processes, such as paper making, cement production, and chemical filtering. Clay is also often used in the manufacture of pipes for smoking tobacco. Until the late 20th century, bentonite clay was widely used as a mold binder in the manufacture of sand castings.

Clay, being relatively impermeable to water, is also used where natural seals are needed, such as in the cores of dams, or as a barrier in landfills against toxic seepage (lining the landfill, preferably in combination with geotextiles).

Recent studies have investigated clay's absorption capacities in various applications, such as the removal of heavy metals from waste water and air purification.

Medicinal Uses

A traditional use of clay as medicine goes back to prehistoric times. An example is Armenian bole, which is used to soothe an upset stomach, similar to the way parrots (and later, humans) in South America originally used it. Kaolin clay and attapulgite have been used as anti-diarrheal medicines.

As a Building Material

Clay is one of the oldest building materials on Earth, among other ancient, naturally-occurring geologic materials such as stone and organic materials like wood. Between one-half and two-thirds of the world's population, in traditional societies as well as developed countries, still live or work in a building made with clay as an essential part of its load-bearing structure. Also a primary ingredient in many natural building techniques, clay is used to create adobe, cob, cordwood, and rammed earth structures and building elements such as wattle and daub, clay plaster, clay render case, clay floors and clay paints and ceramic building material. Clay was used as a mortar in brick chimneys and stone walls where protected from water.

References

- Cane, R.F. (1976). "The origin and formation of oil shale". In Teh Fu Yen; Chilingar, George V. Oil Shale. Amsterdam: Elsevier. pp. 1–12; 56. ISBN 978-0-444-41408-3. Retrieved 5 June 2009.

- Bartis, James T.; LaTourrette, Tom; Dixon, Lloyd; Peterson, D.J.; Cecchine, Gary (2005). "Oil Shale Development in the United States. Prospects and Policy Issues. Prepared for the National Energy Technology Laboratory of the U.S. Department of Energy" (PDF). RAND Corporation. ISBN 978-0-8330-3848-7. Retrieved 29 June

- Carol, Mattusch (2008). Oleson, John Peter, ed. Metalworking and Tools. The Oxford Handbook of Engineering and Technology in the Classical World. Oxford University Press. pp. 418–38 (432). ISBN 978-0-19-518731-1

- Irby-Massie, Georgia L.; Keyser, Paul T. (2002). Greek Science of the Hellenistic Era: A Sourcebook. Routledge. 9.1 "Theophrastos", p.228. ISBN 0-415-23847-1

- Needham, Joseph; Golas, Peter J (1999). Science and Civilisation in China. Cambridge University Press. pp. 186–91. ISBN 978-0-521-58000-7.

- Trench, Richard; Hillman, Ellis (1993). London under London: a subterranean guide (Second ed.). London: John Murray. p. 33. ISBN 0-7195-5288-5.

- Wrigley, EA (1990). Continuity, Chance and Change: The Character of the Industrial Revolution in England. Cambridge University Press. ISBN 0-521-39657-3.

- Speight, James G. (2008). Synthetic Fuels Handbook: Properties, Process, and Performance. McGraw-Hill Professional. pp. 9–10. ISBN 978-0-07-149023-8.

- Rao, P. N. (2007). "Moulding materials". Manufacturing technology: foundry, forming and welding (2 ed.). New Delhi: Tata McGraw-Hill. p. 107. ISBN 978-0-07-463180-5.

- "Lünen – State-of-theArt Ultra Supercritical Steam Power Plant Under Construction" (PDF). Siemens AG. Retrieved 21 July 2014.

- "Gasification Systems 2010 Worldwide Gasification Database". United States Department of Energy. Archived from the original on 1 March 2013. Retrieved March 3, 2013.

- Ken Yin (27 February 2012). "China develops coal-to-olefins projects, which could lead to ethylene self-sufficiency". ICIS Chemical Business. Retrieved 3 March 2013.

- Didi Kirsten Tatlow (5 February 2013). "Worse Than Poisoned Water: Dwindling Water, in China's North" (Blog in edited newspaper). International Herald Tribune. Retrieved 3 March 2013.

- "BP Statistical review of world energy 2012". British Petroleum. Archived from the original (XLS) on 19 June 2012. Retrieved 18 August 2011.

Impacts of Mining on Environment

Mining has adverse effects on the environment, and protective measures need to be taken in regard to this. The environmental effect of mining includes erosion, loss of biodiversity and contamination of ground water and surface water. The topics discussed in the chapter are of great importance to broaden the existing knowledge on the impacts of mining on the environment.

Environmental Impact of Mining

The environmental impact of mining includes erosion, formation of sinkholes, loss of biodiversity, and contamination of soil, groundwater and surface water by chemicals from mining processes. In some cases, additional forest logging is done in the vicinity of mines to increase the gold volume for the sick and room for the storage of the created debris and soil. Besides creating environmental damage, the contamination resulting from leakage of chemicals also affect the health of the local population. Mining companies in some countries are required to follow environmental and rehabilitation codes, ensuring the area mined is returned close to its original state. Some mining methods may have significant environmental and public health effects. Nuss and Eckelman (2014) provide an overview of the life-cycle wide environmental impacts of metals production associated with 62 metals in year 2008.

Acid mine drainage in the Rio Tinto River.

Erosion of exposed hillsides, mine dumps, tailings dams and resultant siltation of drainages, creeks and rivers can significantly impact the surrounding areas, a prime example being the giant Ok Tedi Mine in Papua New Guinea. In areas of wilderness mining may cause destruction and disturbance of ecosystems and habitats, and in areas of farming it may disturb or destroy productive grazing and croplands. In urbanised environments mining may produce noise pollution, dust pollution and visual pollution.

Issues

Water Pollution

Acid mine drainage in Portugal

Mining can have adverse effects on surrounding surface and groundwater if protective measures are not taken. The result can be unnaturally high concentrations of some chemicals, such as arsenic, sulfuric acid, and mercury over a significant area of surface or subsurface. Runoff of mere soil or rock debris -although non-toxic- also devastates the surrounding vegetation. The dumping of the runoff in surface waters or in forests is the worst option here. Submarine tailings disposal is regarded as a better option (if the soil is pumped to a great depth). Mere land storage and refilling of the mine after it has been depleted is even better, if no forests need to be cleared for the storage of the debris. There is potential for massive contamination of the area surrounding mines due to the various chemicals used in the mining process as well as the potentially damaging compounds and metals removed from the ground with the ore. Large amounts of water produced from mine drainage, mine cooling, aqueous extraction and other mining processes increases the potential for these chemicals to contaminate ground and surface water. In well-regulated mines, hydrologists and geologists take careful measurements of water and soil to exclude any type of water contamination that could be caused by the mine's operations. The reducing or eliminating of environmental degradation is enforced in modern American mining by federal and state law, by restricting operators to meet standards for protecting surface and ground water from contamination. This is best done through the use of non-toxic extraction processes as bioleaching. If the project site becomes nonetheless polluted, mitigation techniques such as acid mine drainage (AMD) need to be performed.

The five principal technologies used to monitor and control water flow at mine sites are diversion systems, containment ponds, groundwater pumping systems, subsurface drainage systems, and subsurface barriers. In the case of AMD, contaminated water is generally pumped to a treatment facility that neutralizes the contaminants.

A 2006 review of environmental impact statements found that "water quality predictions made after considering the effects of mitigations largely underestimated actual impacts to groundwater, seeps, and surface water".

Acid Rock Drainage

Heavy Metals

Dissolution and transport of metals and heavy metals by run-off and ground water is another example of environmental problems with mining, such as the Britannia Mine, a former copper mine near Vancouver, British Columbia. Tar Creek, an abandoned mining area in Picher, Oklahoma that is now an Environmental Protection Agency superfund site, also suffers from heavy metal contamination. Water in the mine containing dissolved heavy metals such as lead and cadmium leaked into local groundwater, contaminating it. Long-term storage of tailings and dust can lead to additional problems, as they can be easily blown off site by wind, as occurred at Skouriotissa, an abandoned copper mine in Cyprus.

Effects on Biodiversity

The Ok Tedi River is contaminated by tailings from a nearby mine.

The implantation of a mine is a major habitat modification, and smaller perturbations occur on a larger scale than exploitation site, mine-waste residuals contamination of the environment for example. Adverse effects can be observed long after the end of the mine activity. Destruction or drastic modification of the original site and anthropogenic substances release can have major impact on biodiversity in the area. Destruction of the habitat is the main component of biodiversity losses, but direct poisoning caused by mine-extracted material, and indirect poisoning through food and water, can also affect animals, vegetals and microorganisms. Habitat modification such as pH and temperature modification disturb communities in the area. Endemic species are especially sensitive, since they need very specific environmental conditions. Destruction or slight modification of their habitat puts them at the risk of extinction. Habitats can be damaged when there is not enough terrestrial as well by non-chemicals products, such as large rocks from the mines that are discarded in the surrounding landscape with no concern for impacts on natural habitat.

Concentrations of heavy metals are known to decrease with distance from the mine, and effects on biodiversity follow the same pattern. Impacts can vary greatly depending on mobility and bioavailability of the contaminant: less-mobile molecules will stay inert in the environment while highly mobile molecules will easily move into another compartment or be taken up by organisms. For example, speciation of metals in sediments could modify their bioavailability, and thus their toxicity for aquatic organisms.

Bioaccumulation plays an important role in polluted habitats: mining impacts on biodiversity should be, assuming that concentration levels are not high enough to directly kill exposed organisms, greater on the species on top of the food chain because of this phenomenon.

Adverse mining effects on biodiversity depend to a great extent on the nature of the contaminant, the level of concentration at which it can be found in the environment, and the nature of the ecosystem itself. Some species are quite resistant to anthropogenic disturbances, while some others will completely disappear from the contaminated zone. Time alone does not seem to allow the habitat to recover completely from the contamination. Remediation takes time, and in most of the cases will not enable the recovery of the diversity present before the mining activity.

Aquatic Organisms

The mining industry can impact aquatic biodiversity through different ways. Direct poisoning is the first one, and risks are higher when contaminants are mobile in the sediment or bioavailable in the water. Mine drainage can modify water pH, and it is hard to differentiate direct impact on organisms from impacts caused by pH changes. Effects can nonetheless be observed and proved to be caused by pH modifications. Contaminants can also affect aquatic organisms through physical effects: streams with high concentrations of suspended sediment limit light, thus diminishing algae biomass. Metal oxide deposition can limit biomass by coating algae or their substrate, thereby preventing colonization.

Contaminated Osisko lake in Rouyn-Noranda

Factors that impact communities in acid mine drainage sites vary temporarily and seasonally: temperature, rainfall, pH, salinisation and metal quantity all display variations on the long term, and can heavily affect communities. Changes in pH or temperature can affect metal solubility, and thereby the bioavailable quantity that directly impact organisms. Moreover, contamination persists over time: ninety years after a pyrite mine closure, water pH was still very low and microorganisms populations consisted mainly of acidophil bacteria.

Microorganisms

Algae communities are less diverse in acidic water containing high zinc concentration, and mine drainage stress decrease their primary production. Diatoms community is greatly modified by any chemical change. pH phytoplankton assemblage, and high metal concentration diminishes the abundance of planktonic species. Some diatom species may however grow in high-metal-concentration sediments. In sediments close to the surface, cysts suffer from corrosion and heavy coating. In very polluted conditions, total algae biomass is quite low, and the planktonic diatom community missing. In case of functional complementarity however, it is possible that phytoplankton and zooplankton mass remains stable.

Macroorganisms

Water insect and crustacean communities are modified around a mine, resulting in a low trophic completeness and community being dominated by predators. However, biodiversity of macroinvertebrates can remain high, if sensitive species are replaced with tolerant ones. When diversity is on the contrary reduced, there is sometimes no effect of stream contamination on abundance or biomass, suggesting that tolerant species fulfilling the same function take the place of sensible species in polluted sites. pH diminution in addition to elevated metal concentration can also have adverse effects on macroinvertebrates' behaviour, showing that direct toxicity is not the only issue. Fishes are also affected by pH, temperature variations and chemical concentrations.

Terrestrial Organisms

Vegetation

Soils' texture and water content can be greatly modified in disturbed sites, leading to plants communities changes in the area. Most of the plants have a low concentration tolerance for metals in the soil, but sensitivity differs among species. Grass diversity and total cover is less affected by high contaminant concentration than forbs and shrubs. Mines waste-material rejects or traces due to mining activity can be found in the vicinity of the mine, sometimes pretty far away from the source. Established plants cannot move away from perturbations, and will eventually die if their habitat is contaminated by heavy metals or metalloids at concentration too elevated for their physiology. Some species are more resistant and will survive these levels, and some non-native species that can tolerate these concentrations in the soil, will migrate in the mine surrounding lands to occupy the ecological niche.

Plants can be affected through direct poisoning, for example arsenic soil content reduces bryophyte diversity. Soil acidification through pH diminution by chemical contamination can also lead to a diminished species number. Contaminants can modify or disturb microorganisms, thus modifying nutrient availability, causing a loss of vegetation in the area. Some tree roots avoid the deeper soil layer in order to avoid the contaminated zone, and thus miss anchorage and might be uprooted by the wind when their height and shoot weight increase. In general, root exploration is reduced in contaminated areas compared to non-polluted ones. Even in reclaimed habitats, plant species diversity is lower than in undisturbed areas.

Cultivated crops might be a problem near mines. Most crops can grow on weakly contaminated sites, but yield is generally lower than it would have been in regular growing conditions. Plants also tend to accumulate heavy metals in their aerian organs, possibly leading to human intake through fruits and vegetables. Regular consumptions might lead to health problems caused by long-term metal exposure. Cigarettes made from tobacco growing on contaminated sites might as well have adverse effects on human population, as tobacco tends to accumulate cadmium and zinc in its leaves.

Animals

Malartic mine - Osisko

Habitat destruction is one of the main issue of mining activity. Huge areas of natural habitat are destroyed during mine construction and exploitation, forcing animals to leave the site.

Animals can be poisoned directly by mine products and residuals. Bioaccumulation in the plants or the smaller organisms they eat can also lead to poisoning: horses, goats and sheep are exposed in certain areas to potentially toxic concentration of copper and lead in grass. They are fewer number of ants species in soil containing high copper levels, in the vicinity of a copper mine. If fewer ants are found, chances are great that other organisms leaving in the surrounding landscape are strongly affected as well by this high copper levels, since ants are a good environmental control: they live directly in the soil and are thus pretty sensitive to environmental disruptions.

Microorganisms

Because of their size, microorganisms are extremely sensitive to environmental modification,such as modified pH, temperature changes or chemicals concentration. For example, the presence of arsenic and antimony in soils led to a diminution in total soil bacteria. Moreover, as in water, a small change in the soil pH can provoke the remobilization of contaminants, in addition to the direct impact on pH-sensitive organisms.

Microorganisms have a wide variety of genes among their total population, so there is a greater chance of survival of the species due to the existence of resistance or tolerance genes in some colonies, as long as modifications are not too extreme. Nevertheless, survival in these conditions will imply a big loss of gene diversity, resulting in reduced potential adaptations to subsequent changes. The presence of few developed soil in heavy metal contaminated areas could be a sign of reduced activity by soils microfauna and microflora, indicating a reduced number of individuals or reduced activity. Twenty years after disturbance, even in rehabilitation area, microbial biomass is still greatly reduced compared to undisturbed habitat. Arbuscular mycorrhiza fungi are especially sensitive to the presence of chemicals, and the soil is sometimes so disturbed that they are no longer able to associate with root plants. Some fungi possess however contaminant accumulation capacity, soil cleaning capacity by changing the biodisponibility of contaminants, and can protect plants from damages caused by chemicals. Their presence in contaminated sites could prevent loss of biodiversity due to mine-waste contamination, or allow bioremediation, that is, the removal of undesired chemicals from contaminated soils. On the contrary, some microbes can deteriorate the environment: which mean elevated SO_4 in the water can also increase microbial production of hydrogen sulfide, a toxin for many aquatic plants and organisms.

Effects of Mine Pollution on Humans

Humans are also affected by mining. There are many diseases that can come from the pollutants that are released into the air and water during the mining process. For example, during smelting operations enormous quantities of air pollutants, such as the suspended particulate matter, SO_x, arsenic particles and cadmium are emitted. Metals are usually emitted into the air as particulates.

There are also many occupational health hazards. Most of the miners suffer from various respiratory and skin diseases. Miners working in different types of mines suffer from asbestosis, silicosis, black lung disease etc.

Coal Mining

Deforestation

With open cast mining the overburden, which may be covered in forest, must be removed before the mining can commence. Although the deforestation due to mining may be small compared to the total amount it may lead to species extinction if there is a high level of local endemism.

Mountaintop Removal Mining

Sand Mining

Sand mining and gravel mining creates large pits and fissures in the earth's surface. At times, mining can extend so deeply that it affects ground water, springs, underground wells, and the water table.

Subsidence

Salt mining and salt dome collapsing in Assumption Parish, Louisiana caused the Bayou Corne sinkhole in 2012. As of August 2013, the sinkhole continues to expand.

House in Gladbeck, Germany, with fissures caused by gravity erosion due to mining.

Tailings and Spoil

- Tailings

- Spoil tip

Mitigation

To ensure completion of reclamation, or restoring mine land for future use, many governments and regulatory authorities around the world require that mining companies post a bond to be held in escrow until productivity of reclaimed land has been convincingly demonstrated, although if cleanup procedures are more expensive than the size of the bond, the bond may simply be abandoned. Since 1978 the mining industry has reclaimed more than 2 million acres (8,000 km²) of land in the United States alone. This reclaimed land has renewed vegetation and wildlife in previous mining lands and can even be used for farming and ranching.

Coalworker's Pneumoconiosis

Coal workers' pneumoconiosis (CWP), also known as black lung disease or black lung, is caused by long exposure to coal dust. It is common in coal miners and others who work with coal. It is similar to both silicosis from inhaling silica dust, and to the long-term effects of tobacco smoking. Inhaled coal dust progressively builds up in the lungs and cannot be removed by the body; this leads to inflammation, fibrosis, and in worse cases, necrosis.

Coal workers' pneumoconiosis, severe state, develops after the initial, milder form of the disease known as anthracosis (*anthrac* — coal, carbon). This is often asymptomatic and is found to at least some extent in all urban dwellers due to air pollution. Prolonged exposure to large amounts of coal dust can result in more serious forms of the disease, *simple coal workers' pneumoconiosis* and *complicated coal workers' pneumoconiosis* (or progressive massive fibrosis, or PMF).

More commonly, workers exposed to coal dust develop industrial bronchitis, clinically defined as chronic bronchitis (i.e. productive cough for 3 months per year for at least 2 years) associated with workplace dust exposure. The incidence of industrial bronchitis varies with age, job, exposure, and smoking. In nonsmokers (who are less prone to develop bronchitis than smokers), studies of coal miners have shown a 16% to 17% incidence of industrial bronchitis.

In 2013 CWP resulted in 25,000 deaths down from 29,000 deaths in 1990.

Pathogenesis

Coal dust is not as fibrogenic as in silica dust. Coal dust that enters the lungs can neither be destroyed nor removed by the body. The particles are engulfed by resident alveolar or interstitial macrophages and remain in the lungs, residing in the connective tissue or pulmonary lymph nodes. Coal dust provides a sufficient stimulus for the macrophage to release various products, including enzymes, cytokines, oxygen radicals, and fibroblast growth factors, which are important in the inflammation and fibrosis of CWP. Aggregations of carbon-laden macrophages can be visualised under a microscope as granular, black areas. In serious cases, the lung may grossly appear black. These aggregations can cause inflammation and fibrosis, as well as the formation of nodular lesions within the lungs. The centres of dense lesions may become necrotic due to ischemia, leading to large cavities within the lung.

Appearance

Simple CWP is marked by the presence of 1–2 mm nodular aggregations of anthracotic macrophages, supported by a fine collagen network, within the lungs. Those 1–2 mm in diameter are known as *coal macules*, with larger aggregations known as *coal nodules*. These structures occur most frequently around the initial site of coal dust accumulation — the upper regions of the lungs around respiratory bronchioles. The coal macule is the basic pathological feature of CWP, and has a surrounding area of enlargement of the airspace, known as focal emphysema.

Continued exposure to coal dust following the development of simple CWP may progress to complicated CWP with progressive massive fibrosis (PMF), wherein large masses of dense fibrosis develop, usually in the upper lung zones, measuring greater than 1 cm in diameter, with accompanying decreased lung function. These cases generally require a number of years to develop. Grossly, the lung itself appears blackened. Pathologically, these consist of fibrosis with haphazardly-arranged collagen and many pigment-laden macrophages and abundant free pigment. Radiographically, CWP can appear strikingly similar to silicosis. In simple CWP, small rounded nodules predominate, tending to first appear in the upper lung zones. The nodules may coalesce and form large opacities (>1 cm), characterizing complicated CWP, or PMF.

Diagnosis

There are three basic criteria for the diagnosis of CWP:

1. Chest radiography consistent with CWP

2. An exposure history to coal dust (typically underground coal mining) of sufficient amount and latency

3. Exclusion of alternative diagnoses (mimics of CWP)

Symptoms and pulmonary function testing relate to the degree of respiratory impairment, but are not part of the diagnostic criteria. As noted above, the chest X-ray appearance for CWP can be virtually indistinguishable from silicosis. Chest CT, particularly high-resolution scanning (HRCT), are more sensitive than plain X-ray for detecting the small round opacities.

Epidemiology

In 2013 CWP resulted in 25,000 deaths down from 29,000 deaths in 1990. Between 1970-1974, prevalence of CWP among US coal miners who had worked over 25 years was 32%; the same group saw a prevalence of 9% in 2005-2006.

History

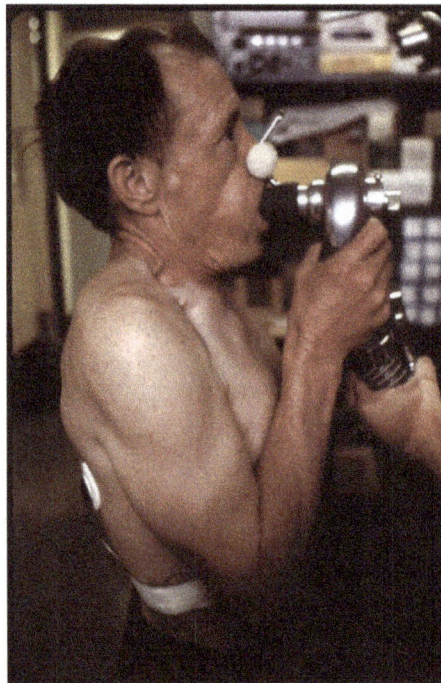

A miner at the Black Lung Laboratory in the Appalachian Regional Hospital in Beckley, West Virginia.

Black lung is actually a set of conditions and until the 1950s its dangers were not well understood. The prevailing view was that silicosis was very serious but it was solely caused by silica and not coal dust. The miners' union, the United Mine Workers of America, realized that rapid mechanization meant drills that produced much more dust, but under John L. Lewis they decided not to raise the black lung issue because it might impede the mechanization that was producing higher productivity and higher wages. Union priorities were to maintain the viability of the long-fought-for welfare and retirement fund, and that required higher outputs of coal. After the death of Lewis, the union dropped its opposition to calling black lung a disease, and realized the financial advantages of a fund for its disabled members.

In the Federal Coal Mine Health and Safety Act of 1969, the US Congress set up standards to reduce dust and created the Black Lung Disability Trust. The mining companies agreed to a clause, by which a ten-year history of mine work, coupled with X-ray or autopsy evidence of severe lung damage, guaranteed compensation. Equally important was a "rate retention" clause that allowed workers with progressive lung disease to transfer to jobs with lower exposure without loss of pay, seniority, or benefits. Financed by a federal tax on coal, the Trust by 2009 had distributed over $44 billion in benefits to miners disabled by the disease and their widows. A miner who spent 25 years in underground coal mines has a 5–10% risk of contracting the disease.

21st Century

After the Federal Coal Mine Health and Safety Act of 1969 became law in the United States, the percentage of American miners suffering from black lung disease decreased by about 90 percent. More recently, however, rates of the disease have been on the rise. The National Institute for Occupational Safety and Health (NIOSH) reported that close to 9 percent of miners with 25 years or more experience tested positive for black lung in 2005–2006, compared with 4 percent in the late 1990s.

New findings have shown that CWP can be a risk for surface coal miners, who are 48% of the workforce. Data from the Coal Workers' Health Surveillance Program of NIOSH, which examined chest X-rays from more than 2,000 miners in 16 US states from 2010-2011, showed that 2% of miners with greater than one year of surface mining experience developed CWP. 0.5% of these miners had PMF. Most of these workers had never worked in an underground mine prior to surface mining. A high proportion of the X-rays suggested that these miners had developed silicosis.

NIOSH, with support from the Mine Safety and Health Administration (MSHA), operates a Mobile Health Screening Program, which travels to mining regions around the United States. Miners who participate in the Program receive health evaluations once every five years, at no cost to themselves. Chest x-rays can detect the early signs of and changes in CWP, often before the miner is aware of any lung problems.

Research

Work to investigate the relationship between respirable dust exposure and coal worker's pneumoconiosis was carried out in Britain by the Institute of Occupational Medicine. This research was known as the Pneumoconiosis Field Research (PFR). The research underpinned the recommendations for more stringent airborne dust standards in British coalmines and the PFR was ultimately used as the basis for many national dust standards around the world.

Acid Mine Drainage

Acid mine drainage, acid and metalliferous drainage (AMD), or acid rock drainage (ARD) refers to the outflow of acidic water from metal mines or coal mines.

Acid rock drainage occurs naturally within some environments as part of the rock weathering process but is exacerbated by large-scale earth disturbances characteristic of mining and other large construction activities, usually within rocks containing an abundance of sulfide minerals. Areas where the earth has been disturbed (e.g. construction sites, subdivisions, and transportation corridors) may create acid rock drainage. In many localities, the liquid that drains from coal stocks, coal handling facilities, coal washeries, and coal waste tips can be highly acidic, and in such cases it is treated as acid rock drainage.

Yellow boy in a stream receiving acid drainage from surface coal mining.

The same type of chemical reactions and processes may occur through the disturbance of acid sulfate soils formed under coastal or estuarine conditions after the last major sea level rise, and constitutes a similar environmental hazard.

Rocks stained by acid mine drainage on Shamokin Creek

Nomenclature

Historically, the acidic discharges from active or abandoned mines were called acid mine drainage, or AMD. The term acid rock drainage, or ARD, was introduced in the 1980s and 1990s to indicate that acidic drainage can originate from sources other than mines. For example, a paper presented in 1991 at a major international conference on this subject was titled: "The Prediction of Acid Rock Drainage - Lessons from the Database" Both AMD and ARD refer to low pH or acidic waters caused by the oxidation of sulfide minerals, though ARD is the more generic name.

In cases where drainage from a mine is not acidic and has dissolved metals or metalloids, or was originally acidic, but has been neutralized along its flow path, then it is described as "Neutral Mine Drainage", "Mining-Influenced Water" or otherwise. None of these other names have gained general acceptance.

Occurrence

In this case, the pyrite has dissolved away yielding a cube shape and residual gold.
This break down is the main driver of acid mine drainage.

Sub-surface mining often progresses below the water table, so water must be constantly pumped out of the mine in order to prevent flooding. When a mine is abandoned, the pumping ceases, and water floods the mine. This introduction of water is the initial step in most acid rock drainage situations. Tailings piles or ponds, mine waste rock dumps, and coal spoils are also an important source of acid mine drainage.

After being exposed to air and water, oxidation of metal sulfides (often pyrite, which is iron-sulfide) within the surrounding rock and overburden generates acidity. Colonies of bacteria and archaea greatly accelerate the decomposition of metal ions, although the reactions also occur in an abiotic environment. These microbes, called extremophiles for their ability to survive in harsh conditions, occur naturally in the rock, but limited water and oxygen supplies usually keep their numbers low. Special extremophiles known as Acidophiles especially favor the low pH levels of abandoned mines. In particular, *Acidithiobacillus ferrooxidans* is a key contributor to pyrite oxidation.

Metal mines may generate highly acidic discharges where the ore is a sulfide mineral or is associated with pyrite. In these cases the predominant metal ion may not be iron but rather zinc, copper, or nickel. The most commonly mined ore of copper, chalcopyrite, is itself a copper-iron-sulfide and occurs with a range of other sulfides. Thus, copper mines are often major culprits of acid mine drainage.

At some mines, acidic drainage is detected within 2–5 years after mining begins, whereas at other mines, it is not detected for several decades. In addition, acidic drainage may be generated for decades or centuries after it is first detected. For this reason, acid mine drainage is considered a serious long-term environmental problem associated with mining.

Chemistry

The chemistry of oxidation of pyrites, the production of ferrous ions and subsequently ferric ions, is very complex, and this complexity has considerably inhibited the design of effective treatment options.

Although a host of chemical processes contribute to acid mine drainage, pyrite oxidation is by far the greatest contributor. A general equation for this process is:

$$2FeS_2(s) + 7O_2(g) + 2H_2O(l) = 2Fe^{2+}(aq) + 4SO_4^{2-}(aq) + 4H^+(aq)$$

The oxidation of the sulfide to sulfate solubilizes the ferrous iron (iron(II)), which is subsequently oxidized to ferric iron (iron(III)):

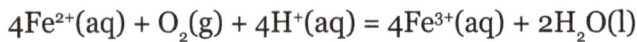

$$4Fe^{2+}(aq) + O_2(g) + 4H^+(aq) = 4Fe^{3+}(aq) + 2H_2O(l)$$

Either of these reactions can occur spontaneously or can be catalyzed by microorganisms that derive energy from the oxidation reaction. The ferric cations produced can also oxidize additional pyrite and reduce into ferrous ions:

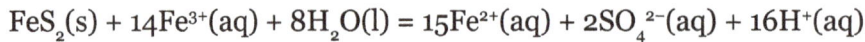

$$FeS_2(s) + 14Fe^{3+}(aq) + 8H_2O(l) = 15Fe^{2+}(aq) + 2SO_4^{2-}(aq) + 16H^+(aq)$$

The net effect of these reactions is to release H$^+$, which lowers the pH and maintains the solubility of the ferric ion.

Effects

Effects on pH

Rio Tinto in Spain.

Water temperatures as high as 47 °C have been measured underground at the Iron Mountain Mine, and the pH can be as low as -3.6.

Organisms which cause acid mine drainage can thrive in waters with pH very close to zero. Negative pH occurs when water evaporates from already acidic pools thereby increasing the concentration of hydrogen ions.

About half of the coal mine discharges in Pennsylvania have pH under 5. However, a significant portion of mine drainage in both the bituminous and anthracite regions of Pennsylvania is alkaline, because limestone in the overburden neutralizes acid before the drainage emanates.

Acid rock drainage has recently been a hindrance to the completion of the construction of Interstate 99 near State College, Pennsylvania. However, this acid rock drainage didn't come from a mine; rather, it was produced by oxidation of pyrite-rich rock which was unearthed during a road cut and then used as filler material in the I-99 construction. A similar situation developed at the Halifax airport in Canada. It is from these and similar experiences that the term acid *rock* drainage has emerged as being preferable to acid *mine* drainage, thereby emphasizing the general nature of the problem.

Yellow Boy

When the pH of acid mine drainage is raised past 3, either through contact with fresh water or neutralizing minerals, previously soluble iron(III) ions precipitate as iron(III) hydroxide, a yellow-orange solid colloquially known as *yellow boy*. Other types of iron precipitates are possible, including iron oxides and oxyhydroxides. All these precipitates can discolor water and smother plant and animal life on the streambed, disrupting stream ecosystems (a specific offense under the Fisheries Act in Canada). The process also produces additional hydrogen ions, which can further decrease pH. In some cases, the concentrations of iron hydroxides in yellow boy are so high, the precipitate can be recovered for commercial use in pigments.

Trace Metal and Semi-metal Contamination

Many acid rock discharges also contain elevated levels of potentially toxic metals, especially nickel and copper with lower levels of a range of trace and semi-metal ions such as lead, arsenic, aluminium, and manganese. The elevated levels of heavy metals can only be dissolved in waters that have a low pH, as is found in the acidic waters produced by pyrite oxidation. In the coal belt around the south Wales valleys in the UK highly acidic nickel-rich discharges from coal stocking sites have proved to be particularly troublesome.

Identification and Prediction

In a mining setting it is leading practice to carry out a geochemical assessment of mine materials during the early stages of a project to determine the potential for AMD. The geochemical assessment aims to map the distribution and variability of key geochemical parameters, acid generating and element leaching characteristics.

The Assessment may Include:

1. Sampling;

2. Static geochemical testwork (e.g. acid-base accounting, sulfur speciation);

3. Kinetic geochemical testwork - Conducting oxygen consumption tests, such as the OxCon, to quantify acidity generation rates

4. Modelling of oxidation, pollutant generation and release; and

5. Modelling of material composition.

Treatment

Oversight

In the United Kingdom, many discharges from abandoned mines are exempt from regulatory control. In such cases the Environment Agency working with partners such as the Coal Authority have provided some innovative solutions, including constructed wetland solutions such as on the River Pelenna in the valley of the River Afan near Port Talbot and the constructed wetland next to the River Neath at Ynysarwed.

Although abandoned underground mines produce most of the acid mine drainage, some recently mined and reclaimed surface mines have produced ARD and have degraded local ground-water and surface-water resources. Acidic water produced at active mines must be neutralized to achieve pH 6-9 before discharge from a mine site to a stream is permitted.

In Canada, work to reduce the effects of acid mine drainage is concentrated under the Mine Environment Neutral Drainage (MEND) program. Total liability from acid rock drainage is estimated to be between $2 billion and $5 billion CAD. Over a period of eight years, MEND claims to have reduced ARD liability by up to $400 million CAD, from an investment of $17.5 million CAD.

Methods

Lime Neutralization

By far, the most commonly used commercial process for treating acid mine drainage is lime precipitation in a high-density sludge (HDS) process. In this application, a slurry of lime is dispersed into a tank containing acid mine drainage and recycled sludge to increase water pH to about 9. At this pH, most toxic metals become insoluble and precipitate, aided by the presence of recycled sludge. Optionally, air may be introduced in this tank to oxidize iron and manganese and assist in their precipitation. The resulting slurry is directed to a sludge-settling vessel, such as a clarifier. In that vessel, clean water will overflow for release, whereas settled metal precipitates (sludge) will be recycled to the acid mine drainage treatment tank, with a sludge-wasting side stream. A number of variations of this process exist, as dictated by the chemistry of ARD, its volume, and other factors. Generally, the products of the HDS process also contain gypsum and unreacted lime, which enhance both its settleability and resistance to re-acidification and metal mobilization.

Less complex variants of this process, such as simple lime neutralization, may involve no more than a lime silo, mixing tank and settling pond. These systems are far less costly to build, but are also less efficient (i.e., longer reaction times are required, and they produce a discharge with higher trace metal concentrations, if present). They would be suitable for relatively small flows or less complex acid mine drainage.

Calcium Silicate Neutralization

A calcium silicate feedstock, made from processed steel slag, can also be used to neutralize active acidity in AMD systems by removing free hydrogen ions from the bulk solution, thereby increasing pH. As the silicate anion captures H^+ ions (raising the pH), it forms monosilicic acid (H_4SiO_4), a neutral solute. Monosilicic acid remains in the bulk solution to play many roles in correcting the adverse effects of acidic conditions. In the bulk solution, the silicate anion is very active in neutralizing H^+ cations in the soil solution. While its mode-of-action is quite different from limestone, the ability of calcium silicate to neutralize acid solutions is equivalent to limestone as evidenced by its CCE value of 90-100% and its relative neutralizing value of 98%.

In the presence of heavy metals, calcium silicate reacts in a different manner than limestone. As limestone raises the pH of the bulk solution, and if heavy metals are present, precipitation of the metal hydroxides (with extremely low solubilities) is normally accelerated and the potential of armoring of limestone particles increases significantly. In the calcium silicate aggregate, as silicic acid species are absorbed onto the metal surface, the development of silica layers (mono- and bi-layers) lead to the formation of colloidal complexes with neutral or negative surface charges. These negatively charged colloids create an electrostatic repulsion with each other (as well as with the negatively charged calcium silicate granules) and the sequestered metal colloids are stabilized and remain in a dispersed state - effectively interrupting metal precipitation and reducing vulnerability of the material to armoring.

Carbonate Neutralization

Generally, limestone or other calcareous strata that could neutralize acid are lacking or deficient at sites that produce acidic rock drainage. Limestone chips may be introduced into sites to create a neutralizing effect. Where limestone has been used, such as at Cwm Rheidol in mid Wales, the positive impact has been much less than anticipated because of the creation of an insoluble calcium sulfate layer on the limestone chips, binding the material and preventing further neutralization.

Ion Exchange

Cation exchange processes have previously been investigated as a potential treatment for acid mine drainage. The principle is that an ion exchange resin can remove potentially toxic metals (cationic resins), or chlorides, sulfates and uranyl sulfate complexes (anionic resins) from mine water. Once the contaminants are adsorbed, the exchange sites on resins must be regenerated, which typically requires acidic and basic reagents and generates a brine that contains the pollutants in a concentrated form. A South African company that won the 2013 IChemE (ww.icheme. org) award for water management and supply (treating AMD) have developed a patented ion-exchange process that treats mine effluents (and AMD) economically.

Constructed Wetlands

Constructed wetlands systems have been proposed during the 1980s to treat acid mine drainage generated by the abandoned coal mines in Eastern Appalachia. Generally, the wetlands receive near-neutral water, after it has been neutralized by (typically) a limestone-based treatment process. Metal precipitation occurs from their oxidation at near-neutral pH, complexation with organic matter, precipitation as carbonates or sulfides. The latter results from sediment-borne anaerobic bacteria capable of reverting sulfate ions into sulfide ions. These sulfide ions can then bind with heavy metal ions, precipitating heavy metals out of solution and effectively reversing the entire process.

The attractiveness of a constructed wetlands solution lies in its relative low cost. They are limited by the metal loads they can deal with (either from high flows or metal concentrations), though current practitioners have succeeded in developing constructed wetlands that treat high volumes and/or highly acidic water (with adequate pre-treatment). Typically, the effluent from constructed wetland receiving near-neutral water will be well-buffered at between 6.5-7.0 and can readily be discharged. Some of metal precipitates retained in sediments are unstable when exposed to oxygen (e.g., copper sulfide or elemental selenium), and it is very important that the wetland sediments remain largely or permanently sub-merged.

An example of an effective constructed wetland is on the Afon Pelena in the River Afan valley above Port Talbot where highly ferruginous discharges from the Whitworth mine have been successfully treated.

Precipitation of Metal Sulfides

Most base metals in acidic solution precipitate in contact with free sulfide, e.g. from H_2S or NaHS. Solid-liquid separation after reaction would produce a base metal-free effluent that can be discharged or further treated to reduce sulfate, and a metal sulfide concentrate with possible economic value.

As an alternative, several researchers have investigated the precipitation of metals using biogenic sulfide. In this process, Sulfate-reducing bacteria oxidize organic matter using sulfate, instead of oxygen. Their metabolic products include bicarbonate, which can neutralize water acidity, and hydrogen sulfide, which forms highly insoluble precipitates with many toxic metals. Although promising, this process has been slow in being adopted for a variety of technical reasons.

Technologies

Many technologies exist for the treatment of AMD from traditional high cost water treatment plants to simple in situ water treatment reagent dosing methods.

Metagenomic Study of Acid Mine Drainage

With the advance of Large-scale sequencing strategies, genomes of microorganisms in the acid mine drainage community are directly sequenced from the environment. The nearly full genomic constructs allows new understanding of the community and able to reconstruct their metabolic pathways. Our knowledge of Acidophiles in acid mine drainage remains rudimentary: we know of many more species associated with ARD than we can establish roles and functions.

Microbes and Drug Discovery

Scientists have recently begun to explore acid mine drainage and mine reclamation sites for unique soil bacteria capable of producing new pharmaceutical leads. Soil microbes have long been a source for effective drugs and new research, such as that conducted at the Center for Pharmaceutical Research and Innovation, suggests these extreme environments to be an untapped source for new discovery.

List of Selected Acid Mine Drainage Sites Worldwide

This list includes both mines producing acid mine drainage and river systems significantly affected by such drainage. It is by no means complete, as worldwide, several thousands of such sites exist.

References

- Derickson, Alan (January 1998). Black lung: anatomy of a public health disaster. Cornell University Press. ISBN 978-0-8014-3186-9.

- Nordstrom, D.K. & Alpers,C. N.: Negative pH, efflorescent mineralogy, and consequences for environmental restoration at the Iron Mountain Superfund site, California PNAS, vol. 96 no. 7, pp 3455–3462, 30 March 1999. Retrieved 4 February 2016.

- "Respiratory Diseases: Occupational Risks". National Institute for Occupational Safety and Health. 21 December 2012. Retrieved 23 March 2015.

- "Overview of acid mine drainage impacts in the West Rand Goldfield". Presentation to DG of DWAF. 2 February 2009. Archived from the original on 2012-03-13. Retrieved 2 July 2014.

- David Falchek (26 December 2012). "Old Forge borehole drains mines for 50 years". The Scranton Times Tribune. Retrieved 18 March 2013.

- Laney AS, Wolfe AL, Petsonk EL, Halldin CN (June 2012). "Pneumoconiosis and advanced occupational lung disease among surface coal miners - 16 states, 2010-2011". MMWR. 61 (23): 431–4. PMID 22695382. Retrieved July 6, 2012.

- Marychurch, Judith; Natalie Stoianoff (4–7 July 2006). "Blurring the Lines of Environmental Responsibility: How Corporate and Public Governance was Circumvented in the Ok Tedi Mining Limited Disaster" (PDF). Australasian Law Teachers Association – Refereed Conference Papers. Victoria University, Melbourne, Victoria, Australia. Archived from the original (PDF) on 7 October 2011. Retrieved 6 December 2011.

- André Sobolewski. "Constructed wetlands for treatment of mine drainage - Coal-generated AMD". Wetlands for the Treatment of Mine Drainage. Retrieved 2010-12-12.

Permissions

All chapters in this book are published with permission under the Creative Commons Attribution Share Alike License or equivalent. Every chapter published in this book has been scrutinized by our experts. Their significance has been extensively debated. The topics covered herein carry significant information for a comprehensive understanding. They may even be implemented as practical applications or may be referred to as a beginning point for further studies.

We would like to thank the editorial team for lending their expertise to make the book truly unique. They have played a crucial role in the development of this book. Without their invaluable contributions this book wouldn't have been possible. They have made vital efforts to compile up to date information on the varied aspects of this subject to make this book a valuable addition to the collection of many professionals and students.

This book was conceptualized with the vision of imparting up-to-date and integrated information in this field. To ensure the same, a matchless editorial board was set up. Every individual on the board went through rigorous rounds of assessment to prove their worth. After which they invested a large part of their time researching and compiling the most relevant data for our readers.

The editorial board has been involved in producing this book since its inception. They have spent rigorous hours researching and exploring the diverse topics which have resulted in the successful publishing of this book. They have passed on their knowledge of decades through this book. To expedite this challenging task, the publisher supported the team at every step. A small team of assistant editors was also appointed to further simplify the editing procedure and attain best results for the readers.

Apart from the editorial board, the designing team has also invested a significant amount of their time in understanding the subject and creating the most relevant covers. They scrutinized every image to scout for the most suitable representation of the subject and create an appropriate cover for the book.

The publishing team has been an ardent support to the editorial, designing and production team. Their endless efforts to recruit the best for this project, has resulted in the accomplishment of this book. They are a veteran in the field of academics and their pool of knowledge is as vast as their experience in printing. Their expertise and guidance has proved useful at every step. Their uncompromising quality standards have made this book an exceptional effort. Their encouragement from time to time has been an inspiration for everyone.

The publisher and the editorial board hope that this book will prove to be a valuable piece of knowledge for students, practitioners and scholars across the globe.

Index

www.ingramcontent.com/pod-product-compliance
Lightning Source LLC
Chambersburg PA
CBHW082038190326
41458CB00010B/3398